O. A. Ortikov
S. S. Rakhimkhodjaev

Design and technology of production of clothing fabrics

O. A. Ortikov
S. S. Rakhimkhodjaev

Design and technology of production of clothing fabrics

with specified properties

ScienciaScripts

Imprint
Any brand names and product names mentioned in this book are subject to trademark, brand or patent protection and are trademarks or registered trademarks of their respective holders. The use of brand names, product names, common names, trade names, product descriptions etc. even without a particular marking in this work is in no way to be construed to mean that such names may be regarded as unrestricted in respect of trademark and brand protection legislation and could thus be used by anyone.

Cover image: www.ingimage.com

This book is a translation from the original published under ISBN 978-620-7-48434-8.

Publisher:
Sciencia Scripts
is a trademark of
Dodo Books Indian Ocean Ltd. and OmniScriptum S.R.L publishing group

120 High Road, East Finchley, London, N2 9ED, United Kingdom
Str. Armeneasca 28/1, office 1, Chisinau MD-2012, Republic of Moldova, Europe
Managing Directors: Ieva Konstantinova, Victoria Ursu
info@omniscriptum.com

Printed at: see last page
ISBN: 978-620-8-36979-8

Contents

Annotation

The work is devoted to design and technology of production of garment fabrics with specified properties. The variants of fabrics with constant and variable raport and number of warp and weft yarn transitions are developed and worked out. At variable raport of warp and weft, and at variable number of warp and weft thread transitions in the fabric: maximum density of warp and weft decreases; geometrical density of warp and weft increases; height of bending wave of warp and weft remains unchanged. At variable warp and weft ratio and at constant value of the number of yarn transitions in the fabric: the maximum density of warp and weft decreases; the geometric density of warp and weft increases; the height of the bending wave in warp and weft remains unchanged. A methodology for calculating the workmanship for each yarn within the fabric rapport is proposed. In all variants, the increase in the number of crossings within the rapport leads to an increase in the yield of warp and weft yarns. The tension of fabric production on the loom depends on the type of weft yarn. With the increase in the linear density of the weft yarn, the yarn yield of the warp increases and that of the weft decreases. The warp yield decreases and weft yield increases when the filling tension of warp yarns is changed from 5sN to 25sN per yarn. The warp yield increases and weft yield decreases when the filling tension of the warp is changed from 5cN to 30cN per thread. The character of the change in the yield obtained analytically and experimentally is identical. The discrepancies in the absolute values of yields are caused by the technological modes, which are not taken into account in the analytical calculations. The selection of parameters and the methodology of determining the values of porosity of the clothing fabric are carried out. The technique of designing the clothing fabric according to the given porosity is developed, where the calculations of thread diameters before and after weaving, fabric density, the coefficient of fabric filling with fibrous material, geometric fabric density, the height of thread bending waves in the fabric, the yarn working out in the fabric are given. Samples of garment fabrics were developed and investigated. In all variants the decrease in the number of crossings within the rapport leads to the decrease in the workmanship of warp and weft yarns in the fabric. At the same weave rapport of the fabric and at reduction of the number of crossings of yarns of one system by another system the porosity increases. With the increase of fabric porosity, the air permeability of clothing fabrics increases. The physical-mechanical, hygienic and consumer properties of clothing fabrics are influenced by the number of thread transitions within the rapport and the type of raw materials used in the weft. With the increase of the number of thread transitions within the rapport, the breaking load on the base and weft, fabric abrasion increases, and the air permeability of the fabric decreases. Capron weft in the fabric increases the breaking load in the weft, breaking elongation in the weft and breathability of the fabric, but reduces the abrasion of the fabric. It is recommended to use garment fabrics which possess good physical-mechanical, hygienic and consumer properties: for variable rapport and variable transition of threads garment fabric of the second variant at the following parameters of fabric structure with rapport on the base and on the weft $R_O = R_y = 6$*, the number of thread transitions within the rapport* t_o = ^= 4,7, P_o =250 threads/dm, P_y =150 threads/dm, T_o =25x2 tex, Tu=15x3 tex; *for constant raport and variable transition of threads garment fabric of the first variant with the following parameters of fabric structure with the rapport on the base and on the weft* $R_O = I_y = 8$*, the number of thread transitions within the rapport* t_o = ^=2,0 , P_o =240 threads / dm., P_y =300 threads / dm, T_o =20 tex, T_y =18,5 x2 tex, *white colour of weft; for constant raport and constant transition of threads the garment fabric of the second variant at the following parameters of fabric structure by rapport on the base and on the weft* R_O = Ry=4, *number of thread transitions within the raport* t_o = í=2, P_o = 258 threads/dm, P_y =190 threads/dm., T_o =25x2 tex, T_y =14,3x3 tex. *Optimal technological parameters of garment fabric production have been developed. The level of warp thread breakage is 0,1 breakage per 1 m. of fabric at warp tension - 16 cN, the value of offset -10 mm. and the position of the scalo above the sternum by 25 mm.*

Abstract: The work is devoted to the project and technology of producing clothing

Keywords: yarn, warp, weft, fabric, rapport, weave, parameters, porosity, workmanship, density, breathability, properties, tension, model.

GENERAL CHARACTERISATION OF WORK

The use of local raw materials for the production of woven fabrics, taking into account the climatic conditions of the region in the products will allow to introduce new technologies, create additional jobs and satisfy the demand for woven products. Therefore, the design and technology of production of clothing fabrics with specified properties is of undoubted interest for weaving production and is actual.

The aim of the research is to investigate the parameters of the structure of clothing fabrics, methodology of designing fabrics with given hygienic properties and optimisation of warp tension on weaving machines.

Objectives of the study.

To address the task at hand, the following is outlined:

- to investigate the structure of fine-patterned weave clothing fabrics;
- to design garment fabrics with specified hygienic properties in terms of porosity and breathability;
- Evaluate the qualities of garment fabrics.
- analytically and experimentally investigate warp tension during the formation of garment fabrics;
- optimise the process of forming garment fabrics;

The object of the study is the structure of clothing fabrics**,** hygienic properties of fabrics, and yarn tension.

The subject of the study is the design and properties of garment fabrics, methods and means of research of warp tension.

Research methodology. To solve the set tasks in the work a complex method of research was used. In the theoretical part, analytical geometry methods were used in the study of tissue structure. In the experimental part of the work the methods of mathematical planning and analysis of results were used. Processing of experimental results was carried out by the method of mathematical statistics. In the technological part the weaving machine "Somet Thema Super Excel" of "Itema" company was used. Standard devices for determination of hygienic properties of fabrics were used, as well as accredited certification centre at TITLP "CentexUz".

The scientific novelty of the study consists in the following:

- variants of fine-patterned weave clothing fabrics with constant and variable rapport and number of warp and weft transitions were developed and investigated;
- The methodology of calculation of workmanship for each thread within

the fabric rapport is proposed, the parameters of garment fabric structure are determined and the garment fabric is designed according to the given porosity;

- regularities of warp tension changes along the width of the existing and modernised dressing of garment fabrics were obtained;
- optimal technological parameters of clothing fabric production on the basis of rotatable central compositional planning of the experiment of the second order have been developed;

Practical results of the research are as follows: samples of fabrics which have good appearance, good air permeability, possess a combination of aesthetic-hygienic and physical-mechanical properties have been developed, which can be recommended to use these fabrics for clothing. Designing of fabrics with given properties of porosity on the basis of local raw materials (cotton) allows to introduce the results of work promptly in the industry and to solve the problem of production of clothing fabrics, which are in great demand in the Republic. Optimisation of technological parameters of clothing fabric production allows to reduce the thread breakage on the warp and to improve the quality of produced fabrics. The methodology of designing fabrics with specified properties, as well as analytical and experimental studies of warp tension across the width of loom threading can be used in the educational process when studying the course of structure and formation of fabrics.

Scientific and practical significance of the research results is characterised by the development of a methodology for designing clothing fabrics with a given porosity. This makes it possible to promptly make filling and production of garment fabrics with minimum costs. Practical significance of the conducted research consists in the creation of a new device for alignment of warp tension across the width of the fabric and optimisation of the process allows to reduce thread breakage and improve the quality of fabrics.

Content of work

The introduction substantiates the relevance of the topic, formulates the purpose, objectives of the study, shows the scientific novelty and practical significance of the work, methodology, research and approbation of the work.

The first chapter provides a review of literature sources devoted to the study of structure, design and formation of fabrics and their properties. Depending on its purpose, fabrics should have appropriate physical and mechanical properties, consumer properties and hygienic properties determined by the type of fibrous material from which the fabric is made, its structure. Physical-mechanical, consumer and hygienic properties of the fabric are characterised by the following main indicators: strength, rigidity, resistance to abrasion, shrinkage after washing, wrinkling, drapability, sliding, lustre, permeability, electrification, etc. The physical-mechanical properties of the fabric are characterised by the following main indicators: strength, rigidity, resistance to abrasion, shrinkage after washing, wrinkling, drapability, sliding, lustre, permeability, electrification, etc. The physical and mechanical properties of technical and special purpose fabrics, in addition, are subject to requirements in accordance with the field of their application. As for household fabrics, especially clothing fabrics, they are subject to hygienic, operational, technological and aesthetic requirements. Such a set of requirements reflects the modern view of the fabric appearance, including its structure. Fabric structure is understood as the mutual arrangement of warp and weft yarns and their connection with each other. The main characteristics of fabric structure are: weave, linear density (diameter) of warp and weft yarns, warp and weft density in the fabric, phase of structure, processing, filling and filling indices, fabric thickness, supporting surface, etc. The main characteristics of fabric structure are: weave, linear density (diameter) of warp and weft yarns, warp and weft density in the fabric, phase of structure, processing, filling and filling indices, fabric thickness, supporting surface, etc. These characteristics can be conditionally divided into two groups - independent and dependent. Independent parameters of fabric structure (basic or initial) - do not depend on other parameters of fabric structure, they are set or accepted during its construction. They include: raw material composition and type of threads and yarns from which the fabric is produced (the structure of thread or yarn, shape and size of cross-section, physical and mechanical properties of yarns depend on the structure and type of fibres; linear density of warp and weft yarns, their diameter; weave of warp and weft yarns in the fabric, determined

by the weave rapport in warp and weft, the number of intersections of weft with warp and warp with weft, overlap shift and the number of yarn layers in the fabric; fabric density in warp and weft. Dependent structure parameters (derivatives) depend on the initial structure parameters of the fabric. For example, fabric thickness depends on the linear density of warp and weft yarns. This group includes: fabric structure phase; warp and weft yarns processing in the fabric; filling and cohesion coefficients; filling coefficient; fabric thickness; supporting surface. All of these 6
the parameters together determine the structure of the fabric and the arrangement of the yarns in it. Also, the structure of the fabric depends on the parameters of fabric formation, such as the size of the backstitch, scalo position, warp and weft tension, and surf force. One of the parameters affecting fabric structure is warp tension during weaving. The warp tension determines the warp deformation during weaving and hence the structure parameters and the type of fabric produced. A distinction is made between warp tensions: per loom cycle; as the warp is triggered on the warp warp; according to the filling width of the weaving machine; in unsteady loom operation (loom start-stop). The filling tension (at shed closing) is necessary to create resistance for the warp yarns when the weft comes to the fabric edge and to ensure a clean shed opening. The filling tension varies depending on the fabric type, being higher for denser (heavy) fabrics and lower for less dense (light) fabrics. Incorrectly chosen filling tension causes violation of the technological process of fabric formation, changes its structure, increases the breakage of warp threads and reduces the efficiency of machine use and the quality of produced fabrics. Decrease of the filling tension causes decrease of warp thread tension at weft surfing, which from cycle to cycle of the machine operation increases the surfing stripe and weaving process becomes impossible due to fabric stuffing. An increase in the filling tension leads to a decrease in the size of the surf stripe, which can lead to overstressing of the warp yarns.

The minimum destructive effect on the warp yarns during weaving can be ensured in such a tension mode, in which the tension variation during the machine cycle is minimal and changes smoothly at the lowest possible filling tension. To regulate (compensate) the warp tension, a movable scalp system is used. Electro-mechanical scalo loading systems are more efficient because they provide for maximum warp tension during the weft surfing period (the shortest time of warp tension) and minimum warp tension in the scoring position and maximum open shed (long-term or the longest time of warp

tension unloading). This allows fabrics with high weft
density (heavy fabrics
) to be produced with
a minimum warp filling tension. The filling tension of the warp yarns must be constant as the warp is worked on the warp. It can be monitored and controlled by yarn tension deviation (with a scalo), perturbation (with a feeler gauge), deviation and perturbation (with a scalo and a feeler gauge). Reverse and electronic warp tension control systems are the most effective. The different warp tension of individual warp yarns along the width of the machine threading is explained by the difference in their physical and mechanical properties, the processes of preparing the warp yarns for weaving in the preparation department, the difference in the width of warp thread picking in the reed and at the fabric edge, as well as the difference in the movement of the wefts. Yarn tension equalisation across the warp width can be achieved by improving yarn quality, by following the parameters of yarn preparation processes, by installing continuous spars, by covering the scal surface with elastic material. In special systems of additional load on the warp yarns during the first weft laying, an additional moment is created on the moving system of the scalo: mechanically - by means of an eccentric; pneumomechanically - by means of a pneumatic cylinder; electromechanically - by means of an electromagnet. The most effective are mechanical and electromechanical systems of reversible action, controlled by a converter from a microprocessor. The value of the backstop is set depending on the type of raw material used for warp and weft yarns, the type of weave of the fabric threads, the density of the fabric in the weft and the design features of the machine. With the increase of the backstop value, the tension of the main threads at the moment of surfing increases, and this changes the structure of the fabric and its properties. Depending on the type of fabric being produced, the scalo moves vertically from the sternum level and this changes the ratio of the upper and lower shed lengths.

In the second chapter the parameters of the structure of fine-patterned weave fabrics are considered. The structure of a fabric is usually understood as the mutual arrangement of warp and weft yarns in the fabric due to their interaction. The forces of interaction between threads in the fabric are created in the process of its formation on the weaving machine and determine the mutual arrangement of threads in the fabric. Mutual arrangement of threads in the fabric, and therefore depends on many factors: type of raw material used; diameters of main and weft threads and their ratios; warp and weft densities and their ratios; type of weave of threads in the fabric; tension of main and weft threads and tension ratios; technological parameters of threading and fabric production. The type of raw materials for the designed fabric is chosen taking into account the purpose of the fabric and the requirements to it. The properties of the yarns used in the warp and weft largely determine the properties of the fabric made of them. A change in the type of raw material in at least one system of yarns in the warp or weft of the fabric has a significant impact on the technological parameters of its production, the structure of the fabric and its properties. The diameters of warp and weft yarns used for fabric production have a significant influence on the technological parameters of fabric production, not its structure and properties. When designing a fabric, yarn diameters are determined depending on the purpose of the fabric and the requirements that 8

The weft yarn diameter increases the breaking load and elongation of the fabric in the weft direction. Increasing the diameter of weft yarns increases the breaking load and elongation of the fabric in the weft direction, the working of warp yarns and decreases the working of weft yarns. Consequently, the ratio of warp and weft yarn diameters has a great influence on the parameters, structure and properties of the fabric. Fabric density of warp and weft, and their ratios have a great influence on the structure and properties of fabrics. The change of fabric density by weft, other things being equal, causes the change of technological parameters of production, structure and properties of fabrics. In particular, an increase in weft density leads to an increase in warp tension and a decrease in warp working and fabric width. The warp and weft densities depend on the diameter of the yarns used and the type of weave of the yarns in the fabric. The maximum possible density of plain weave fabrics (with short overlaps) has a lower value than any other weave (with long overlaps) in the fabric (corduroy, rep, twill, sateen, etc.). Fabrics with the highest possible warp and weft densities are very difficult to weave on the loom. Most of the fabrics produced have a density of either or both yarn systems that is less than the maximum. Therefore, the ratio of the actual fabric density to the maximum density characterises the filling of the fabric with fibrous material, i.e. the tension of the fabric production on the loom. The type of weave has a great influence on the structure and properties of the fabric. In particular, plain weave fabrics (with short overlaps) have a higher breaking load and warp and weft working out than fabrics of other weaves (with long overlaps), and the working out of the fabric (with short overlaps) is accompanied by a high tension. Among the technological parameters that have a significant influence on the structure and properties of the fabric are the tension of the main and weft yarns and their ratios, which change the arrangement of yarns in the fabric, and consequently, the working of yarns in the fabric, the breaking load of the fabric. Another main parameter of threading on the machine is the value of the backstitch, which determines the value of additional warp tension at the moment of forming (surfing) the fabric. As the backstitch increases, the tension of the warp threads at the moment of surf increases, which will lead to changes in the structure of the fabric - density, yarn work in the fabric, breaking load and elongation of the fabric. Consequently, the structure and properties of fabrics can be changed by changing the filling tension and the size of the machine offset. Besides, the structure and properties of the fabrics are influenced by the position of the scalo, the height and depth of the shed, the position of the spar, etc. Here we set the task to study the influence of fabric structure parameters in the production of woven fabrics with fine-patterned weave. Since fine-patterned (combined) weaves have short and long overlaps within the weaving pattern, it is to be expected that the yarns have different stress states both when the fabric is formed on the loom and after the fabric is removed from the loom. On the basis of changing the ratio coefficient of yarn diameters K_d from 0.5 to 2 for a given linear density of warp and weft yarns T_o and T_u, *the* coefficient for cotton yarns S_o and S_u, *the* coefficient of reduction of transverse dimensions of yarns in the fabric η_o and P_u, ***it*** is possible to calculate the diameters of warp d_o and weft d_y, ***the*** bending wave heights of warp h_o and weft h_y, geometrical density of warp yarns l_o and weft yarns l_y, limit and maximum density of fabric on warp P_o and on weft P_y, coefficients determining the order of fabric structure phase on warp K_{ho} and on weft K_{hy}, provided in the first variant when warp yarns are located without gaps $l_o{=}d_o$, in the second variant when weft yarns are located without gaps $l_y{=}d_y$. The calculations were carried out according to the known methodology.

1.The diameter of the thread in the fabric:

warp to weft

$$d_o= 0,03162\, \eta_o\, C_o \sqrt{To} \quad (2.1) \qquad d_y = 0,03162\, \eta_y\, C_y \sqrt{Ty}, \quad (2.2)$$

average thread diameter $d_{cp} = \frac{d_o + d_y}{2}$

2. Yarn diameter in the fabric as a function of the diameter ratio and the average yarn diameter

warp to weft

$$d'_o = \frac{2K_d d_{cp}}{K_d + 1} \quad (2.3) \qquad d'_y = \frac{2d_{cp}}{K_d + 1} \quad (2.4)$$

3. Limit density of the fabric by

warp to weft

$$P_o = \frac{100}{d_o} \quad (2.5) \qquad P_y = \frac{100}{d_y} \quad (2.6)$$

4. Maximum fabric density

warp to weft

$$P'_o = \frac{100}{l_o} \quad (2.7) \qquad P'_y = \frac{100}{l_y}. \quad (2.8)$$

5. Height of the filament bending waves

warp to weft

$$h_o = d_{cp} \cdot K_{ho} \quad (2.9) \qquad h_y = d_{cp} \cdot K_{hy} \quad (2.10)$$

6.Geometric density of short overlap weave fabrics

warp to weft

$$l_o = \sqrt{(d_o + d_y)^2 - h_o^2} \quad (2.11) \qquad l_y = \sqrt{(d_o + d_y)^2 - h_y^2} \quad (2.12)$$

7. Geometric mean density of the fabric for long overlap weaves

on the basis of

$$l_{ocp} = \frac{t_y \cdot \sqrt{(d_o + d_y)^2 - h_o{}^2} + (R_o - t_y) \cdot d_o}{R_o} \quad (2.13)$$

on duck

$$l_{ycp} = \frac{t_o \cdot \sqrt{(d_o + d_y)^2 - h_y{}^2} + (R_y - t_o) \cdot d_y}{R_y} \quad (2.14)$$

where: t_o, t_y - the number of transitions of main threads and, respectively, the number of weft thread transitions from one side of the fabric to the other side of the fabric within the fabric rapport per thread; R_o, R_y - *the* weave rapport of the fabric on the warp and weft.

It follows from the formula that the maximum values of geometric density correspond to the weave ratio of the fabric equal to the number of transitions and these values of geometric density decrease with increasing difference between the weave ratio of the fabric and the number of transitions of warp and weft yarns. For a fine-patterned fabric with a weave pattern on the warp $R_o = 8$ and on the weft $R_y = 8$, with a linear density of warp threads $T_o = 25x2$ tex and with a linear density of weft threads $T_U = 25x2$ tex, the coefficients of the ratio of thread diameters ***Kd*** = 0.5:2 are determined in the variants where the threads are located without gaps on the warp $l_o = d_o$ (Table 2.1) and without gaps in weft $l_y = d_y$ (Table 2.2), warp and weft yarn diameters, height of warp and weft bending wave, geometric density of warp and weft yarns, maximum fabric density in warp and weft, coefficients determining the order of the fabric structure phase.

Table 2.1.

Calculation results for warp yarns of 25x2 tex linear density without gaps in the fabric

Diameter ratio K_d	Diameter of warp yarns d_o, mm	Diameter of weft yarns d_y, mm	Base density limits P_o, n/dm	Bending wave height		Weft geometric density l_y, mm	Maximum weft density P_y, n/dm	Phase order factor, K_{ho}
				h_o, mm	weft h_y, mm			
0,5	0,171	0,343	585	0,483	0,031	0,5131	194,9	1,88
0,6	0,193	0,321	518	0,475	0,039	0,5125	195,1	1,85
0,7	0,212	0,302	472	0,468	0,046	0,5120	195,3	1,82
0,8	0,228	0,286	439	0,460	0,054	0,5112	195,6	1,79
0,9	0,243	0,271	412	0,452	0,062	0,5102	196,0	1,76
1,0	0,257	0,257	389	0,445	0,069	0,5093	196,3	1,73
1,2	0,280	0,234	357	0,432	0,082	0,5074	197,1	1,68
1,4	0,300	0,214	333	0,420	0,094	0,5053	197,9	1,63
1,6	0,316	0,198	317	0,406	0,108	0,5025	199,0	1,58
1,8	0,330	0,184	303	0,393	0,121	0,4996	200,2	1,53
2,0	0,343	0,171	292	0,383	0,131	0,4970	201,2	1,49

Table 2.2.

Results of parameter calculations for linear weft yarn arrangement 25x2 tex density without gaps in the fabric

Diameter ratio K_d	Diameter of warp yarns d_o, mm	Diameter of weft yarns d_y, mm	Duck density P_y, n/dm	Bending wave height		Geometric density on the base l_o, mm	Maximum basis density P_o, n/dm	Phase order coefficient, K_{hV}
				h_o, mm	weft h_y, mm			
0,5	0,171	0,343	292	0,131	0,383	0,4970	201,2	1,49
0,6	0,193	0,321	303	0,121	0,393	0,4996	200,2	1,53
0,7	0,212	0,302	317	0,108	0,406	0,5025	199,0	1,58
0,8	0,228	0,286	333	0,094	0,420	0,5053	197,9	1,63
0,9	0,243	0,271	357	0,082	0,432	0,5074	197,1	1,68
1,0	0,257	0,257	389	0,069	0,445	0,5093	196,3	1,73
1,2	0,280	0,234	412	0,062	0,452	0,5102	196,0	1,76
1,4	0,300	0,214	439	0,054	0,460	0,5112	195,6	1,79
1,6	0,316	0,198	472	0,046	0,468	0,5120	195,3	1,82
1,8	0,330	0,184	518	0,039	0,475	0,5125	195,1	1,85
2,0	0,343	0,171	585	0,031	0,483	0,5131	194,9	1,88

Similarly, for the fine-patterned fabric with weave rapports on warp R_o = 8 and on weft R_y = 8, with linear density of warp and weft threads T_o = T_u = *14* tex and coefficient of ratio of thread diameters K_d = 0.5 + 2 are determined in the variants, where the threads are located without gaps on warp $l_o = d_o$ (Table 2. 3) and without gaps on weft ly = dy (Table 2. 4).3) and without gaps in weft $l_y = d_y$ (Table 2.4), warp and weft yarn diameters, height of warp and weft bending wave, geometric density of warp and weft yarns, maximum fabric density in warp and weft, coefficients determining the order of the fabric structure phase.

Table 2.3.

Results of parameter calculations for warp arrangement
14 tex linear density without gaps in the fabric

Diameter ratio K_d	Diameter of warp yarns d_o, mm	Diameter of weft yarns d_y, mm	Ultimate density on base P_o, n/dm	Bending wave height		Weft geometric density l_y, mm	Maximum weft density P_y, n/dm	Coefficient determining the phase order, K_{ho}
				h_o, mm	Weft h_y, mm			
0,5	0,090	0,180	1111	0,255	0,015	0,2696	371	1,88
0,6	0,101	0,174	990	0,250	0,020	0,2692	372	1,85
0,7	0,111	0,166	901	0,246	0,024	0,2689	372	1,82
0,8	0,120	0,158	833	0,242	0,028	0,2685	372	1,79
0,9	0,128	0,147	781	0,238	0,032	0,2681	373	1,76
1,0	0,135	0,135	741	0,234	0,036	0,2680	373	1,73
1,2	0,147	0,128	680	0,226	0,044	0,2664	375	1,68
1,4	0,158	0,120	633	0,219	0,051	0,2651	377	1,63
1,6	0,166	0,111	602	0,213	0,057	0,2639	378	1,58
1,8	0,174	0,101	575	0,207	0,063	0,2625	381	1,53
2,0	0,180	0,090	556	0,201	0,069	0,2610	383	1,49

Table 2.4.

Calculation results for weft yarns of 14 tex linear density without gaps in the fabric

Diameter ratio K_d	Diameter of warp yarns d_o, mm	Diameter of weft yarns d_y, mm	Duck density limits P_y,n/dm	Height of bending waves		Geometric density on the base l_o, mm	Max-mal base density P_o, n/dm	Phase order determining factor, K_{liv}
				Bases h_o, mm	weft h_y, mm			
0,5	0,090	0,180	556	0,069	0,201	0,2610	383	1,49
0,6	0,101	0,174	575	0,063	0,207	0,2625	381	1,53
0,7	0,111	0,166	602	0,057	0,213	0,2639	378	1,58
0,8	0,120	0,158	633	0,051	0,219	0,2651	377	1,63
0,9	0,128	0,090	680	0,044	0,226	0,2664	375	1,68
1,0	0,135	0,135	741	0,036	0,234	0,2680	373	1,73

1,2	0,147	0,128	781	0,032	0,238	0,2681	373	1,76
1,4	0,158	0,120	833	0,028	0,242	0,2685	372	1,79
1,6	0,166	0,111	901	0,024	0,246	0,2689	372	1,82
1,8	0,174	0,101	990	0,020	0,250	0,2692	372	1,85
2,0	0,180	0,171	1111	0,015	0,255	0,2696	371	1,88

The analysis of Tables 2.1-2.4 shows that when the ratio of yarn diameters is changed from 0.5 to 2: for geometric density on the warp equal to the diameter of the warp yarn, the ultimate density on the warp and the height of the bending wave of the main yarns decrease, and the maximum density on the weft and the height of the bending waves of the weft yarns in the fabric slightly increase; for geometric density on the weft equal to the diameter of the weft yarn, the maximum density on the warp and the height of the bending wave of the main yarns decrease, and the ultimate density on the weft and the height of the bending wave of the weft yarns increase. The comparative analysis of the results of the parameters calculation at the arrangement of warp and weft yarns without gaps in the fabric shows the following: with the increase of the yarn linear density: the limit density of warp and weft, the maximum density of warp and weft decreases; the bending wave height of warp and weft, the geometric density of warp and weft increases. The graphs (Fig.2.1-2.8) of dependence of fabric density on warp and weft, height of bending waves of warp and weft yarns on the coefficient of diameter ratio, with geometrical density on warp equal to the diameter of main thread $l_o = d_o$ and geometrical density on weft equal to the diameter of weft yarn $l_y = d_y$ are also constructed.

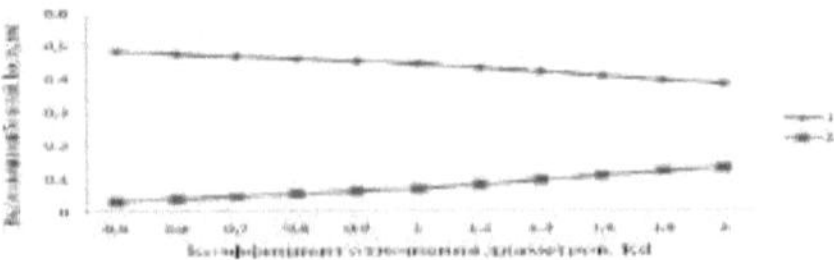

Fig.2.1 The graph of dependence of the bending wave height of warp and weft yarns of linear density 25x2 tex on the diameter ratio, at $l_o = d_o$: where 1- height of bending waves of warp yarns; 2- height of bending waves of weft yarns.

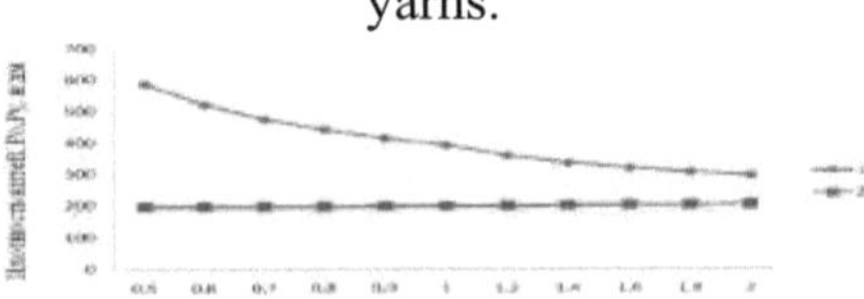

Diameter ratio coefficient, Kd

Fig.2.2: Graph of dependence of fabric density on the basis and weft of linear density 25x2 tex on the coefficient of diameter ratio, at $l_o = d_o$, where:1 - limit density on the basis; 2 - maximum density on the weft.

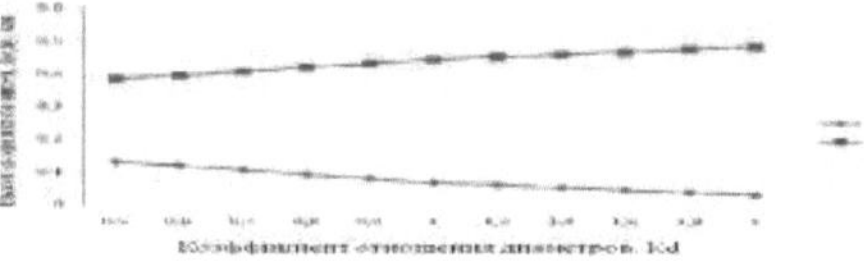

Fig.2.3. Graph of dependence of the height of the bending wave of warp and weft yarns of linear density 25x2 tex on the ratio of diameters, at $l_y = d_y$: where 1- height of the bending waves of warp yarns; 2- height of the bending waves of weft yarns.

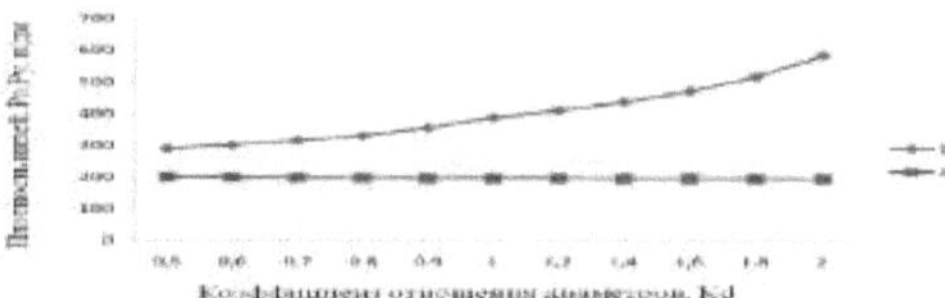

Fig. 2.4. Graph of the dependence of the fabric density on the base and weft of linear density of 25x2 tex on the diameter ratio coefficient, at $l_y = d_y$, where: 1- limit density by weft; 2 - maximum density by warp.

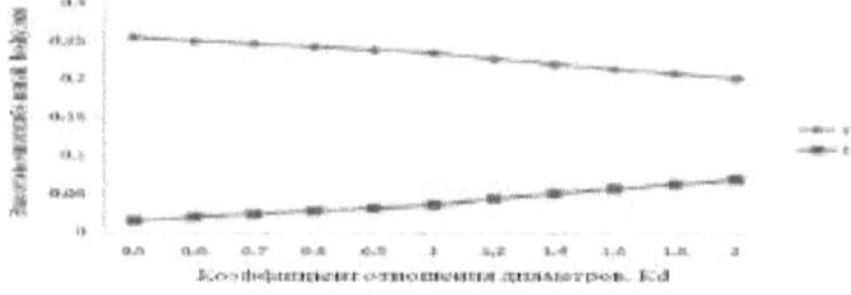

Fig.2.5: Graph of dependence of the height of the bending wave of warp and weft yarns of linear density 14 tex on the ratio of diameters, at $l_o = d_o$: where 1- height of the bending waves of warp yarns; 2- height of the bending waves of weft yarns.

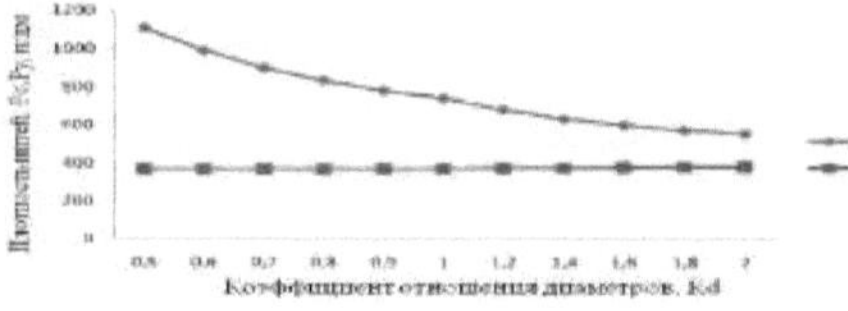

Fig.2.6: Graph of dependence of fabric density by warp and weft of linear density 14 tex on the diameter ratio, at $l_o = d_o$, where: limiting base density; 2 - maximum weft density.

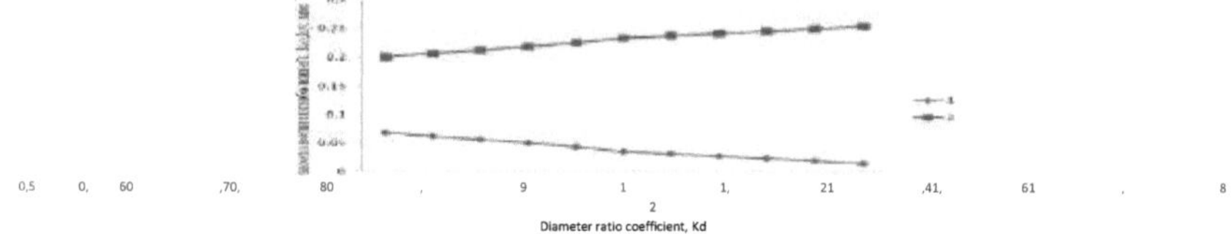

Fig.2.7. Graph of dependence of height of bending wave of warp and weft yarns of linear density 14 tex on the ratio of diameters, at $l_y = d_y$, where 1- height of bending waves of warp yarns; 2- height of bending waves of weft yarns.

Fig.2.8. The graph of dependence of density of fabric on the base and weft of linear

density 14 tex on the coefficient of diameter ratio, at $l_y = d_y$, where $l_y = d_y$:

1- ultimate density in the weft; 2 - maximum density in the base.

Consequently, the ratio of bending waves (warp-weft) shows that the formation of garment fabrics on the *loom*: for the first variant at $l_o = d_o$ (Figs. 2.1, 2.3, 2.5, 2.7) of fabric structure will correspond to the seventh and eighth phase of fabric structure, as ho/hy>1; for the second variant at $l_y = d_y$ (Fig. 2.2, 2.4, 2.6, 2.8) fabric structure will correspond to the second and third phase of fabric structure, as $h_o/h_y < 1$.

Option-1 Option-2

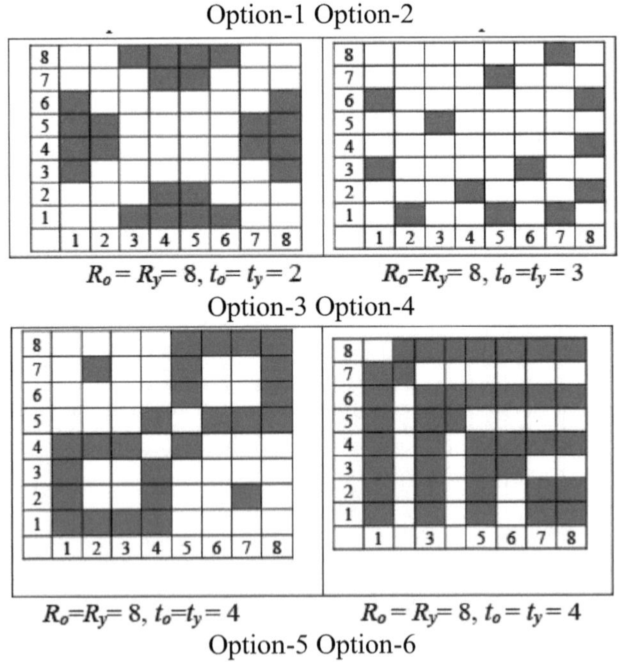

$R_o = R_y = 8$, $t_o = t_y = 2$ $R_o = R_y = 8$, $t_o = t_y = 3$

Option-3 Option-4

$R_o = R_y = 8$, $t_o = t_y = 4$ $R_o = R_y = 8$, $t_o = t_y = 4$

Option-5 Option-6

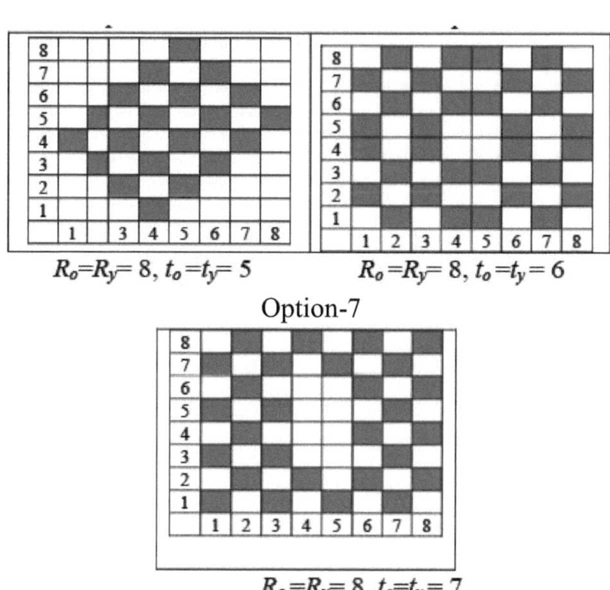

$R_o = R_y = 8,\ t_o = t_y = 5$ $R_o = R_y = 8,\ t_o = t_y = 6$

Option-7

$R_o = R_y = 8,\ t_o = t_y = 7$

Fig.2.9. Variants of fine-patterned weaves of fabrics.

Tables 2.5-2.10 show the effect on geometric and maximum warp and weft densities for a fabric with constant warp rapport $R_o = 8$ and weft rapport $Ry = 8$, variable average values of warp yarn transitions t_o and variable average values of weft yarn transitions t_y in the fabric. And Fig. 2.9 shows the variants of these weaves in the fabric.

Table 2.5.

Geometric and maximum warp and weft densities for fabric $R_o = R_y = 8,\ t_o = t_y = 2$

Tissue structure phase order	Coefficient determining the height of bending waves		Height of bending waves, mm		Geometric density, mm		Maximum density, yarn/dm	
	by wasp new K.	on the duck Khy	by wasp nova, h	weft h	by wasp nove, lo	on duck,[1] at	on the basis of, Ro	by duck, Ru
Marginal	0,27	1,73	0,069	0,445	0,320	0,257	313	389
III	0,5	1,5	0,128	0,386	0,317	0,277	315	361
IV	0,75	1,25	0,193	0,321	0,312	0,293	320	341
V	1	1	0,257	0,257	0,304	0,304	328	328

VI	1,25	0,75	0,321	0,193	0,293	0,312	341	320
VII	1,5	0,5	0,386	0,128	0,277	0,317	361	315
Marginal	1,73	0,27	0,445	0,069	0,257	0,320	389	313

Geometric and maximum warp and weft densities for fabric $R_o = R_y = 8$, $t_o = t_y = 3$

Tissue structure phase order	Coefficient determining the height of bending waves		Height of bending waves, mm		Geometric density, mm		Maximum density, yarn/dm	
	based on Co	duck K1lu	on the basis of, boo	on duck[b] at	by based on, lo	By duck, IY	by on the basis of, Ro	by duck, Ru
Marginal	0,27	1,73	0,069	0,445	0,352	0,257	284	389
III	0,5	1,5	0,128	0,386	0,347	0,275	288	364
IV	0,75	1,25	0,193	0,321	0,339	0,311	295	322
V	1	1	0,257	0,257	0,328	0,328	305	305
VI	1,25	0,75	0,321	0,193	0,311	0,339	322	295
VII	1,5	0,5	0,386	0,128	0,275	0,347	364	288
Marginal	1,73	0,27	0,445	0,069	0,257	0,352	389	284

Table 2.7.

Values of geometric and maximum densities in warp and weft for fabric $Ro=Ry = 8$, $to = ty = 4$

Tissue structure phase order	Coefficient determining the height of bending waves		Height of bending waves, mm		Geometric density, mm		Maximum density, yarn/dm	
	based on Kyaw	on the Kyu duck	by wasp nové, boo	by duck[b] AT	By. wasp nove, lo	on duck, Iv	by on the basis of, Ro	by duck, Ru
Marginal	0,27	1,73	0,069	0,445	0,383	0,257	261	389
III	0,5	1,5	0,128	0,386	0,377	0,281	265	356
IV	0,75	1,25	0,193	0,321	0,367	0,329	272	304
V	1	1	0,257	0,257	0,351	0,351	285	285
VI	1,25	0,75	0,321	0,193	0,329	0,367	304	272
VII	1,5	0,5	0,386	0,128	0,281	0,377	356	256
Marginal	1,73	0,27	0,445	0,069	0,257	0,383	389	261

Values of geometrical and maximum densities in warp and weft for fabric $Ro=Ry = 8$, $to = ty = 5$

Tissue structure phase order	Coefficient determining the height of bending	Height of bending waves, mm	Geometric density, mm	Maximum density, yarn/dm

	waves							
	based on Co	on the Koo duck	on the basis of, boo	by duck by	on the basis of, lo	by duck, IY	by on the basis of, Ro	by duck, Ru
Marginal	0,27	1,73	0,069	0,445	0,415	0,257	241	389
III	0,5	1,5	0,128	0,386	0,408	0,287	245	348
IV	0,75	1,25	0,193	0,321	0,394	0,347	254	288
V	1	1	0,257	0,257	0,375	0,375	267	267
VI	1,25	0,75	0,321	0,193	0,347	0,394	288	254
VII	1,5	0,5	0,386	0,128	0,287	0,408	348	245
Marginal	1,73	0,27	0,445	0,069	0,257	0,415	389	241

Table 2.9.

Values of geometrical and maximum densities on warp and weft for fabric $Ro=Ry= 8$, $to= ty= 6$

Tissue structure phase order	Coefficient determining the height of bending waves		Height of bending waves, mm		Geometric density, mm		Maximum density, yarn/dm	
	based on Co	on the Kyu duck	on the basis of, boo	by duck[b] AT	on the basis of, lo	on duck, lv	by on the basis of, Ro	by duck, Ru
Marginal	0,27	1,73	0,069	0,445	0,446	0,257	224	389
III	0,5	1,5	0,128	0,386	0,438	0,293	228	341
IV	0,75	1,25	0,193	0,321	0,422	0,365	237	274
V	1	1	0,257	0,257	0,398	0,398	251	251
VI	1,25	0,75	0,321	0,193	0,365	0,422	274	237
VII	1,5	0,5	0,386	0,128	0,293	0,438	341	228
Marginal	1,73	0,27	0,445	0,069	0,257	0,446	389	224

Geometric and maximum warp and weft densities for fabric $R_o = R_y = 8$, $t_o = t_y = 7$

Tissue structure phase order	Coefficient determining the height of bending waves		Height of bending waves, mm		Geometric density, mm		Maximum density, yarn/dm	
	based on Kyo	on Kyu's duck	by wasp Nové, Yo.	by duck[a] at	on the basis of, lo	By duck, IY	by on the basis of, Ro	by duck, Ru
Marginal	0,27	1,73	0,069	0,445	0,478	0,257	204	389
III	0,5	1,5	0,128	0,386	0,468	0,300	214	333
IV	0,75	1,25	0,193	0,321	0,450	0,383	222	261
V	1	1	0,257	0,257	0,422	0,422	237	237

VI	1,25	0,75	0,321	0,193	0,383	0,450	261	222
VII	1,5	0,5	0,386	0,128	0,300	0,468	333	214
Marginal	1,73	0,27	0,445	0,069	0,257	0,478	389	204

Comparative analysis of the results of calculation of geometric and maximum warp and weft densities with the number of yarn transitions from 2 to 7 for a fabric with a rapport $R_o = R_y = 8$ shows that with increasing number of yarn transitions within the rapport: the maximum warp and weft densities decrease; the geometric warp and weft densities increase; the height of the warp and weft bending wave remains unchanged. Tables 2.11 to 2.14 show the effect on geometric and maximum warp and weft densities for fabrics with variable warp and weft rapports in fabrics from 2 to 5, and average values of warp and weft yarn transitions in fabrics.

Table 2.11.

Values of geometric and maximum warp and weft densities for 1/4 twill weave

Tissue structure phase order Twill U<	Coefficient determining the height of bending waves		Height of bending waves, mm		Geometric density, mm		Maximum density, yarn/dm	
	based on Kyo	on Kyu's duck	at the core, Yo.	on duck[fl] AT	on the basis of, 1o	on duck, lv	by on the basis of, Ro	by duck, Ru
Marginal	0,27	1,73	0,069	0,445	0,358	0,257	279	389
III	0,5	1,5	0,128	0,386	0,353	0,290	283	345
IV	0,75	1,25	0,193	0,321	0,345	0,315	299	317
V	1	1	0,257	0,257	0,332	0,332	301	301
VI	1,25	0,75	0,321	0,193	0,315	0,345	317	299
VII	1,5	0,5	0,386	0,128	0,290	0,353	345	283
Marginal	1,73	0,27	0,445	0,069	0,257	0,358	389	279

Table 2.12.

Geometric and maximum warp and weft densities for twill weave 1/3

Tissue structure phase order Twill 1/3	Coefficient determining the height of bending waves		Height of bending waves, mm		Geometric density, mm		Maximum density, yarn/dm	
	based on Co	On duck K1lu	by wasp nové, boo	on duck[b] at	on the basis of, 1o	By duck, IY	by on the basis of, Ro	by duck, Ru
Marginal	0,27	1,73	0,069	0,445	0,383	0,257	261	389

III	0,5	1,5	0,128	0,386	0,378	0,298	265	336
IV	0,75	1,25	0,193	0,321	0,367	0,329	273	304
V	1	1	0,257	0,257	0,351	0,351	285	285
VI	1,25	0,75	0,321	0,193	0,329	0,367	304	273
VII	1,5	0,5	0,386	0,128	0,298	0,378	336	265
Marginal	1,73	0,27	0,445	0,069	0,257	0,383	389	261

Table 2.13

Values of geometric and maximum warp and weft densities for twill weave 1/2

Tissue structure phase order Twill %	Coefficient determining the height of bending waves		Height of bending waves, mm		Geometric density, mm		Maximum density, yarn/dm	
	based on *Kho*	on the duck *Khy*	by wasp new, *ho*	By. duck *hy*	On the basis, *lo*	On the duck, *ly*	by on the basis of, *Po*	by duck, *Ru*
Marginal	0,27	1,73	0,069	0,445	0,425	0,257	253	389
III	0,5	1,5	0,128	0,386	0,418	0,312	239	321
IV	0,75	1,25	0,193	0,321	0,403	0,353	248	283
V	1	1	0,257	0,257	0,382	0,382	262	262
VI	1,25	0,75	0,321	0,193	0,353	0,403	283	248
VII	1,5	0,5	0,386	0,128	0,312	0,418	321	239
Marginal	1,73	0,27	0,445	0,069	0,257	0,425	389	253

Values of geometric and maximum warp and weft densities for plain weave 1/1 fabrics

Tissue structure phase order Canvas *1/1*	Coefficient determining the height of bending waves		Height of bending waves, mm		Geometric density, mm		Maximum density, yarn/dm	
	based on *Kho*	on the duck *Khy*	by wasp nova, *hoo*	By. duck *hy*	on the basis of, *lo*	on duck, *ly*	by on the basis of, *Po*	by duck, *Ru*
Marginal	0,27	1,73	0,069	0,445	0,509	0,257	197	*389*
III	0,5	1,5	0,128	0,386	0,498	0,339	205	295
IV	0,75	1,25	0,193	0,321	0,476	0,401	210	249
V	1	1	0,257	0,257	0,445	0,445	225	225
VI	1,25	0,75	0,321	0,193	0,401	0,476	249	210
VII	1,5	0,5	0,386	0,128	0,339	0,498	295	205
Marginal	1,73	0,27	0,445	0,069	0,257	0,509	389	197

From the analysis of Tables 2.11-2.14, it follows that at variable warp and weft counts and at a variable number of warp and weft transitions in the fabric: the maximum warp and weft densities decrease; the geometric warp and weft densities increase; the warp and weft bending wave heights remain unchanged. With variable raveling (Table 2.5, 2.11, 2.12, 2.13, 2.14) in warp and weft and with a constant value of the number of yarn transitions in the fabric: the maximum density in warp and weft decreases; the geometric density in warp and weft increases; the bending wave height in warp and weft remains unchanged. The ratio of the actual density to the maximum density is characterised by the fibre content of the fabric. The fibre fill factor takes into account the fabric density, the linear yarn density and the weave of the fabric. Fill factor

on the basis of: $a_o = (L_o - L_T) / L_o \cdot 100\%$ (2.17)

on the duck: $a_y = (L_y - B_T) / L_y \cdot 100\ \%$ (2.18)

The latter, in addition to the weave rapport, takes into account the number of thread transitions of one system relative to another system, which determine the number of transitions of each thread in the fabric rapport of the fibre filling factor and warp and weft yarn processing in the fabric. It is expedient to determine the number of transitions of each thread in the fabric of fine-patterned weaves and, on the basis of this index, to determine the fibre filling ratio and the warp and weft yarn processing in the fabric.

Table 2.15.

Influence of the number of yarn crossings on the filling factor

№	Parameters determining the structure of the tissue				Fill factor		
	Ro	Ry	to	ty	on the basis of Kno	on Knu's duck	Fabrics KT
1.	8	8	2	2	0,62	0,86	0,53
2.	8	8	3	3	0,70	0,92	0,64
3.	8	8	4	4	0,78	0,99	0,77
4.	8	8	4	4	0,78	0,99	0,77
5.	8	8	5	5	0,86	1,0	0,91
6.	8	8	6	6	0,94	1,1	1,0
7.	8	8	7	7	1,0	1,2	1,2

The analysis of Table 15 shows that for a fabric with a report $Ro = Ry = 8$ in warp and weft, when the number of thread transitions changes from 2 to 7, the filling factor in warp, weft and fabric increases. Therefore, the most intensive process of fabric production on the weaving machine takes place in the seventh variant: at filling factor on warp equal to ***Kno*** = *1*.0; filling factor on weft equal to $K_{nu} = 1.2$; fabric filling factor equal to $K_t = 1.2$. Yield of yarns in the fabric is one of the main parameters, which allows to estimate in the first approximation the conditions of fabric production on the machine. The consumption of raw materials is determined by the yield. Yield is influenced by the type of raw material, linear yarn density, yarn cross-sectional shape, weave, warp and weft density, order of the construction phase, threading and weaving parameters on the weaving machine. The difference between the lengths of the threads before weaving and in the weave is called work-in.

Base yield $ao = (Lo - LT) / Lo \cdot 100\%$ (2.17)

Weft labour rate $ay = (Ly - BT) / Ly \cdot 100\ \%$ (2.18)

where: Lo, Ly - warp and weft lengths before weaving; LT, BT - length and width of the fabric.

Theoretically, the yarn yield in the fabric can be determined from the mutual arrangement of yarns in the fabric (Fig. 2.10). In this case, the following assumptions are made: the shape and cross-sectional area of the yarns along the entire length of the fabric are constant, the distance between the centres of the yarns of one system in the places of their intersection with the yarns of another system and in the overlaps are proportional to the filling factor of the corresponding system.

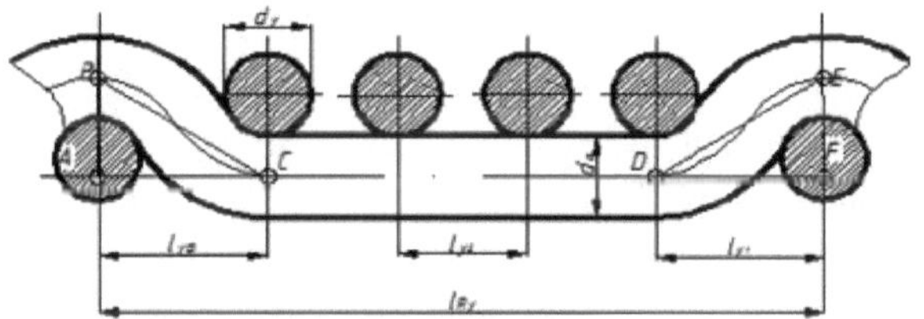

Fig.2.10. Thread arrangement in the fabric

The length of one warp thread in (Lo) is determined (Fig.10) by the length of the broken line BCDE, and the length of the fabric by the straight line ACDF.

Base yield $ao = (L_{BCDE} - L_{ACDF}) / L_{BCDE} \cdot 100$

$ao = (BC+CD+DE-AC-CD-DF)/BC+CD+DE100 =$

$= BC+DE-AC-DE / BC+CD+DE100$

where: BC, DE - areas where the warp yarns cross the weft yarns

$$BC = DE = \sqrt{AC^2 + AB^2} = \sqrt{l_{y\phi}^2 + h_o^2}$$

$l_y f$- actual distance between the centres of the weft yarns at the places of intersection of their main yarns, i.e. taking into account the order of the fabric structure phase and the actual weft density of the fabric

$$l_{y\phi} = l_y / K_{Hy} = (\sqrt{(d_o + d_y)^2 - hy^2}) / K_{Hy} \quad (2.19)$$

The length of the rectilinear part of the thread is equal to the sum of the distances between the threads утка $CD = l_{y2}(R_y - t_{ocp}) = (R_y - t_{ocp})\, d_y / K_{Hy}$

where: l_{y2} is the distance between weft yarns at the long main overlap t_{ocp}, t_{ycp}

Substituting the obtained expressions, we determine the warp yarn work in the fabric

$$a_o= [t_{ocp} (\sqrt{l_{y\phi}^2 + h_o^2} - l_{y\phi}) / t_{ocp} (\sqrt{l_{y\phi}^2 + h_o^2}) + (R_y - t_{ocp})\, d_y/K_{Hy}]\ 100 \qquad (2.20)$$

Similarly determine the weft seam allowance in a fabric

$$a_o= [t_{ycp} (\sqrt{l_{o\phi}^2 + h_y^2} - l_{o\phi}) / t_{ycp} (\sqrt{l_{o\phi}^2 + h_y^2}) + (R_o - t_{ycp})\, d_o/K_{Ho}] \cdot 100 \qquad (2.21)$$

where: K_{Ho}, K_{Hy} - coefficients of filling of fabric with fibrous material in warp and weft.

For fine-patterned weave fabrics caused by short and long overlaps in one rapport and having equally intertwining yarns of each weave motif, the average yarn working is determined according to the formulas.

$$a_o = \frac{100}{R_o} \sum_{i=1}^{n} \frac{t_o(\sqrt{l_{y\phi}^2 + h_o^2} - l_{y\phi}}{t_o\sqrt{l_{y\phi}^2 + h_o^2} + (R_y - t_o)\frac{d_y}{K_{Hy}}} \qquad (2.22)$$

For warp threads:

$$a_y = \frac{100}{R_y} \sum_{i=1}^{n} \frac{t_y(\sqrt{l_{o\phi}^2 + h_y^2} - l_{o\phi}}{t_y\sqrt{l_{o\phi}^2 + h_y^2} + (R_o - t_y)\frac{d_o}{K_{Ho}}} \qquad (2.23)$$

For the weft threads:

where: t_o , t_y - respectively *the* number of warp and weft intersections of equally intertwined yarns within the limits of the fabric rapport for each weave motif; l_{uf}, l_{of} - the actual distance between the centres of the yarns (weft-basic) at the places of their intersection for each weave motif; h_o, h_y - height of the bending wave of the warp and weft yarns respectively; R_o, R_y - weave ratio of the fabric; ***n*** - number of equally intertwined yarns of each weave motif, the sum of these numbers is equal to the weave ratio of the fabric; $(R_y - t_o)\frac{d_y}{K_{Hy}}$, $(R_o - t_y)\frac{d_o}{K_{Ho}}$ - length of the straight part of the yarn within the weave ratio.

The production of fabric variants of fine patterned weaves was carried out on a Thema "Somet Super Excel" weaving machine in the laboratory of the department "Textile Weave Technology". For the fine patterned fabric with the weave pattern on the warp $R_o=12$ and on the weft $R_y=12$, the number in the dressing 12 remise, the reed number N = 60 teeth/dm, the number of threads picked in the reed tooth - 4 threads, the density on the warp 250 n/dm. and on the weft 150 n/dm., linear density of main threads 25x2 tex, while linear density of weft threads varied from 15 tex to 75 tex, warp tension varied for single thread from 5 to 25 cN, for weft thread from 5 to 25 cN. Fig. 2.11 shows the variants of fine-patterned weaves, taking into account the warp and weft weave rapports, the number of crossings of yarns of one system by another system, and the fibre fill factor in warp and weft. Table 2.16 according to Fig. 2.11. presents the fabric parameters for twelve weave variants, and the calculation of the yarn working was carried out by formulae (2.22), (2.23). For convenience in the calculation the intersections are labelled with indices, which can be multiplied by two, e.g.: The intersection t_{o1} and t_{y1} means that this thread has two transitions, i.e. to1=2, $ty1=2$. For t_{o2} – 4 , t_{y2} = 4, to3=6, t_{y3} = 6, etc. Intersections and the number of equal-weave are determined according to Figure 2.11. Calculations carried out (at T_{y}= 45 tex) according to formulas (2.20) and (2.21) show that with the same fabric rapport and the same intersection (variants II, III weave), the average weft and warp yarns are the same.

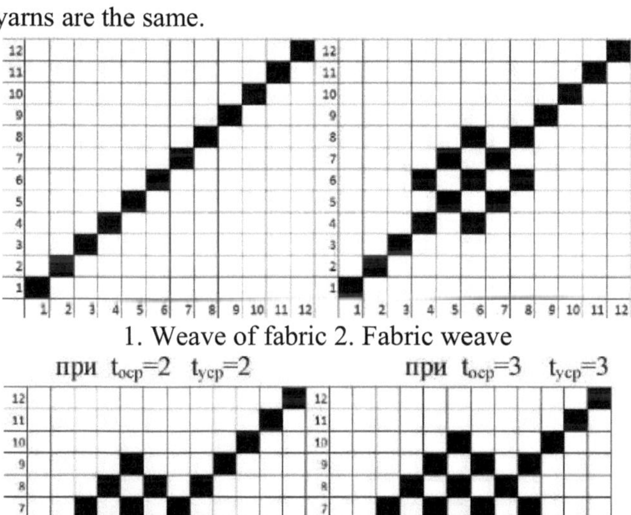

1. Weave of fabric 2. Fabric weave

при $t_{оср}$=2 $t_{уср}$=2 при $t_{оср}$=3 $t_{уср}$=3

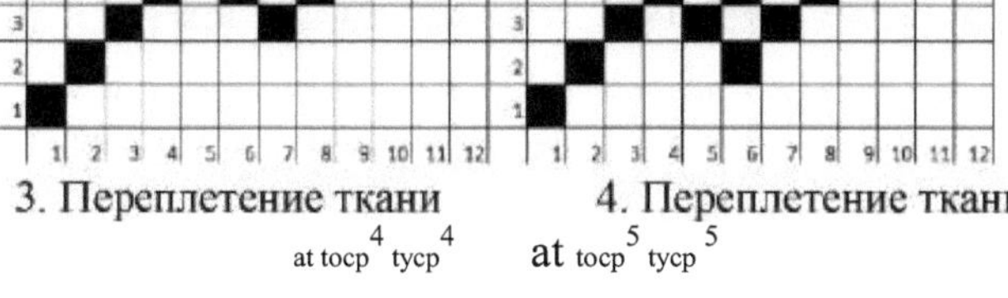

3. Переплетение ткани 4. Переплетение ткани

at $tocp^{4}$ $tycp^{4}$ at $tocp^{5}$ $tycp^{5}$

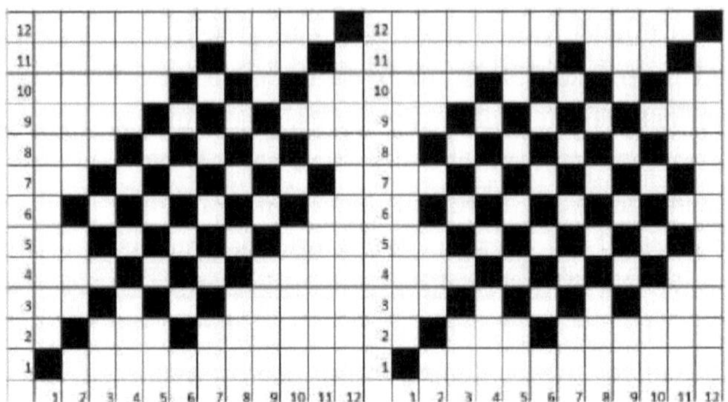

5. Weaving of fabric 6. Fabric weave

at tocp^{6} tycp^{6} at tocp^{7} tycp^{7}

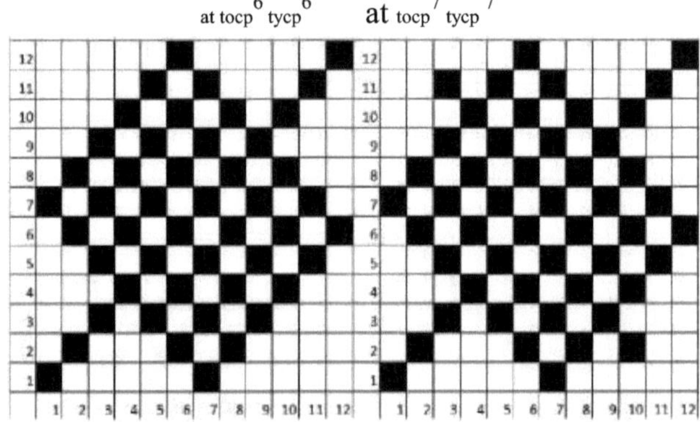

7. Переплетение ткани 8. Переплетение ткани

at tocp^{8} tycp^{8} at tocp^{9} tycp^{9}

9. Weave of the fabric 10. Fabric weave

at tocp 10 tycp 10PI tocp 11 tycp 11

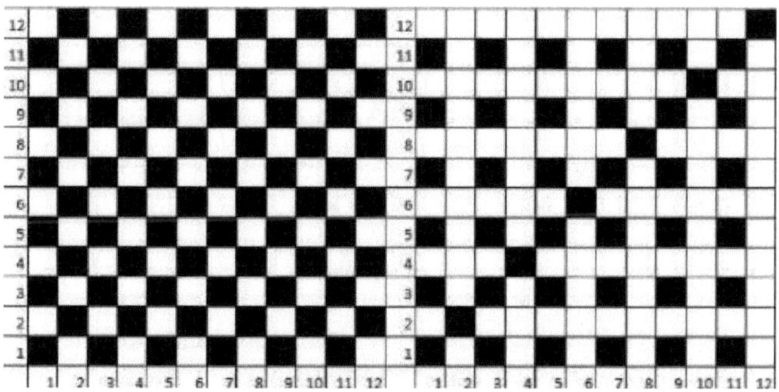

11. Weaving of fabric12 . Fabric weave

n^p and tocp=^{12,} tycp=^{12} n^p and tocp=^{7,} tycp=^{7}

Fig.2.11. Variants of fine-patterned weaves of fabrics.

Table 2.16

Parameters and results of calculations of yarn processing of fine-patterned fabrics according to the proposed methodology

Weave options	1	2	3	4	5	6	7	8	9	10	11	12
t01/ni	2/12	2/7	2/5	2/5	2/2	2/2	-	-	-	-	-	2/6
Wn2	-	4/4	4/2	-	4/2	-	4/2	4/2	-	-	-	-
Ypz	-	6/1	6/5	6/4	6/4	6/2	6/2	-	6/2	-	-	-
Yp4	-	-	-	8/2	8/2	8/6	8/4	8/2	-	8/1	8/1	-
t05/n5	-	-	-	10/1	10/2	10/2	10/2	10/6	10/6	10/4	10/4	-
t06/n	-	-	-	-	-	-	12/2	12/2	12/4	12/7	12/7	12/6
tyl/ni	2/12	2/7	2/5	2/3	2/2	2/2	-	-	-	-	-	2/6
ty2/n2	-	4/4	4/2	4/2	4/2	4/2	4/2	4/2	-	-	-	-
ty3/n3	-	6/1	6/5	6/5	6/2	-	6/2	-	6/2	-	-	-
ty4/n4	-	-	-	8/2	8/2	8/4	8/4	8/2	-	8/1	-	-
ty5/n	-	-	-	-	10/2	10/4	10/2	10/6	10/6	10/4	-	-
ty6/n6	-	-	-	-	-	-	12/7	12/2	12/4	12/7	-	-
a 01,%	1,4	1,6	1,8	2,1	2,3	-	-	-	-	-	-	2,6
a 02,%	-	2,1	3,2	-	4,8	-	5,1	6,2	-	-	-	-
a 03,%	-	2,9	4,3	4,4	5	5,1	6,5	-	8,3	-	-	-
a 04,%	-	-	-	4,6	6,2	6,3	7,7	9	-	10,4	11,5	-
a 05,%	-	-	-	5	7,2	7,3	8,7	9,8	10,6	11,6	12,3	-
a 06,%	-	-	-	-	-	-	9,5	10,9	11,4	12,8	13,4	8,8
^{a} osr,%	1,4	2,2	3,1	4,0	5,1	6,2	7,5	9,0	10,1	11,6	12,4	5,7
au1,%	1,5	1,7	1,9	2,1	2,4	2,6	-	-	-	-	-	2,6
au2,%	-	2,2	3,4	4,4	4,5	5,8	5,9	6,6	-	-	-	-
au3,%	-	3,3	4,9	6,2	6,5	-	6,2	-	9,6	-	-	-
au4,%	-	-	-	7,3	7,4	9,4	9,5	10,2	-	12,3	13,4	-
au5,%	-	-	-	-	8,2	10,6	10,8	11,2	12,2	13,7	14,6	-
au6,%	-	-	-	-	-	-	10,9	12,4	13,6	14,5	15,2	10,4
ausr,%	1,5	2,4	3,4	5,0	5,8	7,1	8,6	10,1	11,8	13,5	14,4	6,5

For variants 6 and 12 at $R_o = R_y = 12$:
in the first case $t_{01} = 2$ at n=2, $t_{03} = 6$ at n=2, $t_{04} = 8$ at n=6, $t_{05} = 10$ at n=2, tyi=2 at n=2, ty2=4 at n=2,ty4=8 at n=4, ty5=10 at n=4, average uranium yield on the base a_o = 6,2 %, on the weft a_y= 7,1 % ;
in the second case $t_{01} = t_{y1} = 2$ at n=6, $t_{06} = t_{y6} = 12$ at n=6, average yield on the base a_o = 5,7 % , on the weft a_y = 6,5 % .
As can be seen the number of intersections in the rapport is the same, but in the second case these intersections have extreme values $t_{oi} = t_{yi}$ and $t_{o6} = t_{y6}$, and in the first case have average values t_{o3}, t_{y2}, $t_{o4} = t_{y4}$ and $t_{o5} = t_{y5}$ in the interval from **ti to t6** , which leads to a decrease in warp yield by 8 % and weft yield by 9 %. In addition, it can be seen that an increase in the number of intersections (in all variants) within the rapport leads to an increase in the yield of warp and weft yarns. Similarly, the calculations of yarn yield for the average number of intersections in warp and weft were carried out by formulae (2.20) and (2.21) according to Fig. 2.11, the results of which are presented in Table 2.17.

Table 2.17.

Parameters and results of calculations of yarn processing of fine-patterned fabrics using the existing methodology

Weave options	Ro	Ry	losr	lusr	KNo	Knu	losr, mm	lusr, mm	ao,%	ay,%
1	12	12	2,0	2,0	0,807	0,467	0,522	1,092	1,4	1,5
2	12	12	3,0	3,0	0,862	0,502	0,479	1,016	2,3	2,5
3	12	12	4,0	4,0	0,919	0,537	0,449	0,950	3,2	3,6
4	12	12	5,0	5,0	0,972	0,572	0,425	0,892	4,2	4,7
5	12	12	6,0	6,0	1,027	0,607	0,402	0,840	5,3	6,0
6	12	12	7,0	7,0	1,082	0,642	0,382	0,794	6,5	7,3
7	12	12	8,0	8,0	1,137	0,676	0,363	0,754	7,7	8,8
8	12	12	9,0	9,0	1,192	0,741	0,346	0,717	9,1	10,3
9	12	12	10,0	10,0	1,248	0,746	0,331	0,684	10,3	11,9
10	12	12	11,0	11,0	1,303	0,781	0,317	0,653	11,6	13,5
11	12	12	12,0	12,0	1,36	0,816	0,304	0,625	13,0	15,2
12	12	12	7,0	7,0	1,082	0,642	0,382	0,794	6,5	7,3

The analysis of Table 2.17 shows that with the same rapport and average number of warp t_{ocp} and weft t_{ycp} crossings in the fabric (weave variants 6,12), the average warp and weft yarns are the same. Though it can be seen from Table 16 that the values of variants 6 and 12 are different, due to different variation in the intersection values (**t**). Tables 2.16 and 2.17 show that as the intersections increase with the same rapport, the values of fill factor (FF) and yarn yield increase. Table 2.18 shows the influence of the linear density of the weft on the filling coefficient and on the yarn work in the fabric. In the general case, with the change of the linear density of the weft from T_y = 15 tex to T_u = 75 tex, the tension of fabric production increases, the yield of warp yarns increases, and the yield of weft yarns decreases.

Table 2.18.

Influence of the linear density of the weft on the yarn work in the fabric

№	Linear density of weft yarn, tex,	Fill factor		Yarn yield, %	
		at the core, KNO.	on the duck, Knu.	at the core, *and* oh.	on the duck, ow.
1	15	0,919	0,474	3,2	7,9
2	30	1,01	0,568	5,1	7,5
3	45	1,082	0,642	6,5	7,3
4	60	1,144	0,703	8,3	6,9
5	75	1,190	0,757	9,2	6,3

Table 2.19

Influence of warp filling tension on yarn development in fabrics

№	Filling tension of warp yarns cN/thread	Filling tension of weft yarns cN/thread	Yarn yield %	
			at the core, *and* oh.	on the duck, ow.
1	5	15	7,5	6,1
2	10	15	6,9	6,7
3	15	15	6,6	7,6
4	20	15	6,2	8,2
5	25	15	5,7	8,7

Table 2.20.

Influence of the filling tension of the weft on yarn development in fabrics

№	Filling tension of weft yarns cN/thread	Filling tension of warp yarns cN/thread	Yarn yield %	
			on the basis of	on duck
1	5	15	5,7	8,7
2	10	15	6,0	8,1
3	15	15	6,6	7,6
4	20	15	6,9	6,9
5	25	15	7,6	6,4

It follows from Table 2.19 that with increasing the filling tension from 5 cN. to 25 cN. per single warp yarn, all other conditions being equal: the warp yield decreases by 30 %, and the weft yield increases by 40 %. At constant filling tension of the warp yarns (Table 2.20) and increasing the filling tension of the weft yarn from 5 cN. to 25 cN. the warp yield increases by 30

%, and the weft yield decreases by 36 %. Tables 2.21-2.22 present the numerical characteristics of the weft and warp yarn yields in the fabric and their errors, which were determined according to the known methodology.

Table 2.21.

Numerical characteristics of weft yarns in fabrics

Options	Yarn processing values					
	Average value of U	Disper-sion8^2 (U)	Mean square deviation S (U)	Coefficient of variation C(U)	Absolute error of mean value E(U)	Relative error of mean value b(Y)
1	2,0	0,007	0,083	4,2	0,1	5,2
2	3,1	0,007	0,083	2,7	0,1	3,3
3	3,9	0,007	0,083	2,1	0,1	2,6
4	4,9	0,026	0,161	3,3	0,2	4,1
5	6,2	0,059	0,242	3,9	0,3	4,8
6	7,6	0,102	0,320	4,2	0,4	5,2
7	8,8	0,102	0,320	3,6	0,4	4,5
8	10,4	0,160	0,400	3,8	0,5	4,7
9	12,1	0,230	0,480	4,0	0,6	5,0
10	13,2	0,230	0,480	3,6	0,6	4,5
11	15,2	0,290	0,560	3,7	0,7	4,6

Table 2.22

Numerical characteristics of warp yield in fabrics

Options	Yarn processing values					
	Average value of U	Dispersion S^2 (U)	Mean square deviation S (Y)	Coefficient of variation C(U)	Absolute error of mean value E(U)	Relative error of the mean value of b(U)
1	1,8	0,007	0,083	4,5	0,1	5,6
2	2,6	0,007	0,083	3,2	0,1	4,0
3	3,3	0,026	0,161	4,8	0,2	5,9
4	4,1	0,026	0,161	3,9	0,2	4,8
5	5,1	0,059	0,242	4,7	0,3	5,8
6	6,6	0,102	0,320	4,8	0,4	5,9
7	8,0	0,059	0,242	3,0	0,3	3,7
8	9,3	0,160	0,400	4,3	0,5	5,3
9	11,0	0,230	0,480	4,4	0,6	5,5
10	12,1	0,160	0,400	3,3	0,5	4,1
11	13,4	0,290	0,560	4,2	0,7	5,2

Comparing the results of Tables 2.21, 2.22 and 2.17, it can be noted that the character of the yield change in all variants is identical, i.e. the yield is influenced by the number of crossings in the fabric rapport and the type of raw materials used in the weft. However, there are large differences in the absolute values of yarn counts, which is caused by the fact that the yield calculated by the formulas (2.20), (2.21), (2.22), (2.23) does not take into account the technological modes of fabric production (warp and weft yarn tension, the size of the gap, the size of the shed).

In the third chapter the design of clothing fabrics with specified properties and their quality assessment is carried out. Clothing serves to regulate the heat output of the human body, creating around the body an artificial, regulated temperature environment, relatively independent of the direct influences of the external environment. In addition, clothing protects the body from mechanical damage and thus contributes to the preservation of health. Clothing replaces, therefore, the natural protective cover that is missing in the human body. Clothing is mainly made of fabrics. The material for fabrics is plant fibres (flax, hemp, cotton, etc.), animal fibres (wool, silk) and artificial (artificial silk) fibres. Clothing in general consists of several layers of fabrics, of different thicknesses and of different nature, depending on the season. According to the requirements of hygiene clothing should be: homogeneous in the sense of the structure of the fabrics that make up its composition, ie, have the same initial between themselves physical properties, appropriate cut, appropriate weight, corresponding to external environmental conditions and the state of the body, ie, the temperature of the ambient air, humidity and its movement, the radiant heat of the sun and the well-being of the body; corresponding to the work performed by a person. The hygienic properties of clothing depend on the properties of the materials from which the clothing is created, i.e. the hygienic properties of fabrics. Hygienic properties of fabrics depend on the properties of the source material (fibre) and the fabric manufacturing technique. Elasticity, moisture capacity and wearability depend on the properties of the source material. Air content (air permeability), water capacity, evaporation, thermal conductivity and heat transfer depend on the processing method. The same results can be obtained from wool, cotton paper and linen if they are processed in such a way that they have the same hygienic properties. Studies of the hygienic properties of fabrics were carried out in the following sequence: studies of the physical properties of fabrics - thickness, weight, specific gravity, porosity; studies of the relation of fabric to air - air permeability; designing fabrics according to a given porosity. Fabric thickness is of great importance in the study and evaluation of hygienic properties of fabric. A number of its properties - air permeability, heat transfer, etc. - depend on the fabric thickness. Differences

in the thickness of fabrics are caused by the desire to give a certain strength to the material, requirements depending on the household purpose of the fabric, the desire to achieve the best heat conservation, etc. The thickness of one and the same fabric in one piece is not the same: it is subject to fluctuations depending on the method of fabric manufacturing. Under the influence of different conditions, the thickness of the fabric changes to one side or the other. When wetting with water, the thickness of the fabric increases or decreases (up to 30%) regardless of the material (wool, linen, cotton). Wetting with oils does not change the thickness of fabrics. Placing the fabric in a space saturated with moisture causes an increase in its thickness. The thickness of the fabric also changes under the influence of wear (decrease), boiling (increase), the action of steam (in disinfection - up to 115 °). Changes in the thickness of fabrics depend on the properties of fibre and fabric (hygroscopicity, wettability, etc.), as well as on the method of manufacture. Changes in thickness cannot but affect other properties of the fabric (thermal conductivity, air permeability, etc.), so that hygienic requirements in this respect are reduced to its possible invariability from various factors and not to the detriment of other properties. The thickness of fabrics can be measured with a special thickness gauge or determined by the formula $T_t = d_o + d_y$.
For unequal degrees of curvature, the thickness of the tissue is
$T_T = 2d_y + d_0$ or $T_T = 2d_o + d_y$. The thickness gauge is designed according to the principle of a spring scale with an arrow and a dial. At the bottom of the instrument box is a spring-loaded rod with a platform resting on the table of the instrument foot. The fabric is laid between this pad and the table, for which purpose the rod with the pad is lifted with the help of the lever on the top of the device and then lowered after laying the fabric. The thickness of the fabric is counted on the dial by the arrow 1 minute after laying the fabric. The accuracy of the arrow is 0.01mm. The fabric consists of a mixture of fibres with air enclosed by the interlacing of these fibres. The volumetric weight of the fabric, i.e. the weight of 1 cm^3 , depends on the amount of the dense substance composing it. The volumetric weight of the fabric depends on the method of weaving and the thickness of the fabric. The hygienic significance of the volumetric weight of a fabric is that the lower the volumetric weight, the more air is contained in the fabric, the looser it is and therefore the easier it is for air to pass through. The volumetric weight of a fabric is determined by re-weighing several fabric samples of known area. The samples are cut on a metal template with a size of 10 x 10 *cm.* Then divide the average weight of the fabric by its area (100 cm^2) and obtain the

weight of 1 cm^2 at natural thickness; then calculate the weight of 1 cm^2 at a thickness of 1 *cm*. The volumetric weight (γ_τ) of the fabric can be expressed by the formula

$$\gamma_T = \frac{10 \cdot в}{n \cdot m}$$

where ***c*** is the weight of 1 cm^2 of tissue, *n-area of* tissue, ***m-thickness***.
Or fabric volumetric weight (γ_τ), calculated from the weight of one metre^2 of fabric and its thickness

$$\gamma_T = \frac{M}{1000 \cdot m} \qquad (3.1)$$

where: M - fabric weight of 1 m^2 in gr; m - fabric thickness in mm.
By comparing the volume weight of the fabric (γ_τ) with the specific gravity of the material in the fibre (γ_β), it is possible to calculate the proportion of the fabric volume filled with material in the fibre. The specific gravity of fibres of different origins is quite close to each other: the specific gravity of cotton is 1.363 mg/mm^3 ; the specific gravity of silk is 1.326 mg/mm^3 ; the specific gravity of wool is 1.296 mg/mm^3 . Therefore, the specific gravity of all fibres can be assumed to be 1.3 mg/mm^3 . Total porosity is calculated as the percentage of fabric volume not filled with fibrous material

$$R_S = 100 \cdot (1 - \frac{\gamma_T}{\gamma_B}) \qquad (3.2)$$

Hygienic properties of fabrics are also assessed by their air permeability, as it affects the carbon dioxide content under the garment; the higher the air permeability of the fabrics that make up the garment, the less carbon dioxide is contained under the garment, i.e. the more hygienic the fabrics are. The breathability of clothing depends on the following factors: the permeability of the fabrics to air; the thickness of the garment; the area of the body (on the chest and more closed parts of the body, ventilation conditions are more difficult, while on the arms under the sleeves they are more favourable); and the cut of the garment. Hence, breathability is the main hygienic property of fabrics. The air permeability of a fabric refers to the physical property that air under some, even slight, pressure passes through the pores of the fabric. Research of air permeability of fabrics and determination of its value is carried out on special devices, the essence of which is reduced to the fact that the tested fabric is passed through the injection or suction of air. During the experiment the following data are taken into account - the volume of air passed through the fabric; the duration of the passage of this volume of air; the area of the fabric through which the air passes; the pressure under which the air passes. Designing of hygienic properties of fabrics according to the given porosity is carried out. Initial data for designing, the porosity of fabric R_s, weave, phase of fabric structure, filling coefficient by warp or weft, linear density of yarn by warp and weft, coefficient of ratio of densities or diameters of yarns, coefficients of change of yarn sizes in fabric are set.

Firstly, the calculated yarn diameter before weaving is determined

$$d_{cp}=0{,}0316C\sqrt{\frac{T_o+T_y}{2}} \qquad (3.3)$$

Determine the density of the fabric on the warp

$$P_o=\frac{100(K_d+1)K_{Ho}}{d_{cp}(K_d\eta_{oe}+\eta_{ye})\sqrt{4-K_{ho}^2}} \qquad (3.4)$$

The weft density of the fabric is expressed through the maximum possible weft density and the unknown fibre fill factor of the fabric.

$$P_y=P_{ymax}\cdot K_{Hy}=\frac{100(Kd+1)K_{Hy}}{d_{cp}(K_d\eta_{oe}+\eta_{ye})\sqrt{4-K_{hy}^2}} \qquad (3.5)$$

Filling factor of the weft fabric with fibrous material is determined from the ratio of fabric porosity

$$R_s=100-d_{oe}\cdot P_o-d_{ye}\cdot P_y+0{,}01d_{oe}\cdot d_{ye}\cdot P_o\cdot P_y \qquad (3.6)$$

Where: $d_{oe}=d_o\cdot\eta_{oe}$, $d_{ye}=d_y\cdot\eta_{ye}$

Theoretical yarn count in warp fabrics

$$a_o=\frac{L_o-L_{TO}}{L_o}\cdot 100\% \qquad (3.7)$$

$$L_o=\sqrt{l_{y\phi}^2+h_o^2}; \qquad (3.8)$$

$$L_{To}=l_{y\phi}=100/P_y; \qquad (3.9)$$

$$h_o=\frac{d_{oe}+d_{ye}}{2}\cdot K_{ho}; \qquad (3.10)$$

Yarn processing in weft fabrics

$$a_y=\frac{L_y-L_{Ty}}{L_y}\cdot 100\% \qquad (3.11)$$

$$L_y=\sqrt{l_{o\phi}^2+h_y^2}; \qquad (3.12)$$

$$L_{Ty}=l_{o\phi}=100/P_o; \qquad (3.13)$$

$$h_y=\frac{d_{cp}(\eta_{oe}+\eta_{ye})}{2}\cdot K_{hy}; \qquad (3.14)$$

Given the task to design a fabric with porosity $R_s = 38 \pm 1$, the linear density of the warp and weft $T_o = 25x2$, $T_y = 45tex$, the coefficient of **C** yarn is 1.25, the coefficient of the ratio of warp and weft diameters before weaving

$$K_d = \frac{d_o}{d_y} = \frac{50}{45} = 1{,}1,$$

weave of the fabric according to Figure 3.1, according to technical requirements the fabric of VI order of phase structure, i.e. fabric of high density on warp and $K_{ho} = 1{,}2$ and $K_{hy} = 0{,}8$; KHO= 0,85, and the coefficient of filling of fabric on weft is determined by the given porosity of fabric, the coefficient of change of diameters of yarns in fabrics

$\eta_{oz} = 1{,}1$, $\eta_{yz} = 1{,}1$, $\eta_{oe} = 0{,}8$, $\eta_{ye} = 0{,}8$.

1.Determine the calculated yarn diameter before weaving by formula (3.3)

$$d_o = d_y = d_{cp} = 0{,}0316 \cdot 1{,}25 \cdot \sqrt{\frac{50+45}{2}} = 0{,}272мм$$

2.Estimated yarn diameters taking into account their dimensional variation in warp and weft fabrics

$$d_{oz} = d_o \cdot \eta_{oz} = 0{,}272 \cdot 1{,}1 = 0{,}299мм$$

$$d_{yz} = d_y \cdot \eta_{yz} = 0{,}272 \cdot 1{,}1 = 0{,}299мм$$

$$d_{oe} = d_o \cdot \eta_{oe} = 0{,}272 \cdot 0{,}8 = 0{,}218мм$$

$$d_{ye} = d_y \cdot \eta_{ye} = 0{,}272 \cdot 0{,}8 = 0{,}218мм$$

3.Determine by (4) the density of fabric on the base

$$P_o = \frac{100(1{,}1+1) \cdot 0{,}85}{0{,}272 \cdot (1 \cdot 1{,}1 + 0{,}8)\sqrt{4-1{,}2^2}} = 246н/дм$$

4.Let us determine by (3.5) the weft density of the fabric through the maximum possible weft density and the unknown value of the fibre filling factor of the fabric

$$P_y = \frac{100(0{,}9+1) \cdot K_{Hy}}{0{,}272 \cdot (0{,}9 \cdot 0{,}8 + 1{,}1)\sqrt{4-0{,}8^2}} = 208K_{Hy},$$

5.Determine the coefficient of filling of fabric by weft with fibrous material by formula (3.6)

$$38{,}5 = 100 - 0{,}299 \cdot 246 - 0{,}299 \cdot 208 \cdot K_{Hy} + 0{,}01 \cdot 0{,}299 \cdot 0{,}299 \cdot 246 \cdot 208 \cdot K_{Hy}$$

From where $38{,}5 = 26{,}5 - 16{,}5 \cdot K_{Hy}$

$$K_{Hy} = \frac{12}{16{,}5} = 0{,}727$$

6.The obtained coefficient provides high tissue connectivity and is within the normal range.

Weft density of fabric after inserting K_{Hy} into formula (3.5)

$P_y = 208\text{-} K_{Hy} = 208\text{-}0.727 = 150 nits / dm$

7.Working up the warp threads in the fabric

$$L_o = \sqrt{0{,}6667^2 - 0{,}261^2} = 0{,}716\text{мм}$$

$$l_{y\phi} = l_{To} = \frac{100}{150} = 0{,}667\text{мм}$$

$$h_o = \frac{0{,}272(0{,}8+0{,}8)}{2} \cdot 1{,}2 = 0{,}261\text{мм};$$

$$a_o = \frac{0{,}716 - 0{,}6667}{0{,}716} \cdot 100 = 6{,}9\%;$$

8. Finishing the weft yarns in the fabric

$$L_y = \sqrt{0{,}407^2 + 0{,}174^2} = 0{,}443\text{мм}, \qquad l_{o\phi} = L_{Ty} = \frac{100}{484} = 0{,}407\text{мм}$$

$$h_y = \frac{0{,}272(0{,}8+0{,}8)}{2} \cdot 0{,}8 = 0{,}174\text{мм}; \qquad a_y = \frac{0{,}443 - 0{,}407}{0{,}443} \cdot 100 = 8{,}1\%$$

The fine patterned fabric was produced on Thema "Somet Super Excel" machine in the laboratory of the Department of Weaving with weave ratio R_o = 12 in warp and R_y = 12 in weft, number of remizas in the dressing 12, reed number N = 60 teeth/dmm, the number of threads picked in the reed tooth - 4 threads, density on the warp 250 n/dm. and on the weft 150 n/dm., linear density of main threads 25x2 tex, linear density of weft threads 45 tex. Figs. 3.1-3.4 show the variants of fine-patterned weaves, taking into account the weave patterns in warp and weft, the number of crossings of yarns of one system by another system, the fibre filling factor in warp and weft.

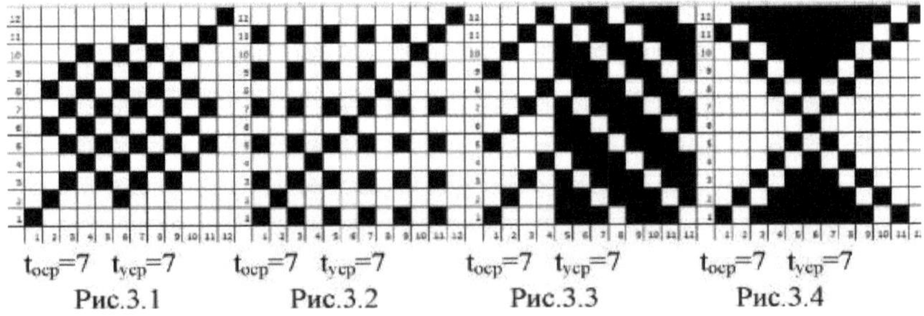

t_{ocp}=7 t_{ycp}=7 Рис.3.1 | t_{ocp}=7 t_{ycp}=7 Рис.3.2 | t_{ocp}=7 t_{ycp}=7 Рис.3.3 | t_{ocp}=7 t_{ycp}=7 Рис.3.4

Tables 3.1 and 3.2 present the numerical characteristics of the warp and weft yarns in the fabric and their errors, which are determined by formulae (2.26) to (2.31).

Table 3.1.

Numerical characteristics of warp yield in fabrics

№	Fabric	Yarn processing values

	sample options	Average value of U	E" s persia S^2 (U)	Mean square deviationZ (Y)	Coefficient of variation and C(U)	Absolute error mean value E(U)	Relative error of the mean value of b(Y)
1	I	6,0	0,04	0,2	3,3	0,3	4,2
2	II	4,2	0,02	0,14	3,4	0,3	4,2
3	III	5,5	0,02	0,14	2,5	0,2	3,1
4	IV	4,5	0,03	0,17	3,8	0,2	4,7

Table 3.2.

Numerical characteristics of weft yarns in fabrics

№	Fabric sample options	Values y threading					
		Average value of U	Dispersion S^2 (U)	Mean square deviationZ (Y)	Coefficient of variation of C(U)	Absolute error mean value E(U)	Relative errorcr unit value b(U)
1	I	7,0	0,07	0,27	3,8	0,3	4,7
2	II	5,3	0,04	0,2	3,8	0,3	4,7
3	III	6,5	0,06	0,25	3,8	0,3	4,7
4	IV	5,5	0,05	0,22	4,0	0,3	4,9

The analysis of tables 3.1 and 3.2 (variants 1 and 2 of fabric samples) shows that at the same fabric rapport and the same thread intersections (Fig. 3.1 and Fig. 3.2), the average yields of warp and weft threads are not the same. For variant 1 (Fig.3.1) - at $R_o=R_y=$ 12, $t_{o1}=t_{y1}=2$ at **n** = 2, $t_{o3}=t_{y3}=2$ at **n** = 2 , $t_{o4}=t_{y4}$ = 8 at **n** = 6, $t_{o5}=t_{y5}=10$ at **n** = 2 average yield of warp a_o = 6,0 %, of weft a_u = 7,0 %.

For variant 2 (Fig.3.2) -$t_{o1}=t_{y1}=2$, at **n** = 6 and $t_{o6}=t_{y6}=12$ at **n** = 6 average yield on the base a_o = 4,2%, on the weft a_u = 5,3%.

As can be seen the number of intersections in the rapport is the same, but in the second case the intersections have extreme values $t_{o1}=t_{y1}$ and $t_{o6}=t_{y6}$, and in the first case the middle values are $t_{o4}=t_{y4}$ И t_{o5} = t_{y5} in the interval from t_1 to t_6, which leads to reduction of average warp yield by 30% and weft yield by 24%. In addition, Table 3.1 and Table 3.2 (variants 3.1-3.4 of fabric samples) show that the decrease in the number of thread crossings within the rapport leads to a decrease in the yield of warp and weft yarns in the fabric.

Table 3.3.

Numerical characteristics of fabric porosity

№	Fabric sample options	Fabric porosity values, %					
		Average value of U	ДисPersia S^2 (U)	Mean square deviation S (U)	Coefficient of variation C(U)	Absolute error mean E(U)	Relative error of mean value b(Y)
1	I	38	0,8	0,9	2,4	1,0	3,0
2	II	53	1,0	1,0	2,0	2,0	3,5
3	III	35	0,8	0,9	2,6	1,0	3,3
4	IV	47	1,1	1,05	2,4	2,0	3,0

Table 3.4.

Numerical characteristics of fabric breathability

№	Fabric sample options	Fabric breathability values, ($slg' cm^2 s$)					
		Average value of U	ДисPersia S^2 (U)	Mean square deviation eS (Y)	Coefficient of variation C(U)	Absolute error mean value E(U)	Relative error of mean b(Y) value
1	I	32	0,6	0,8	2,4	1,0	3,0
2	II	99	2,8	1,7	1,7	4,0	2,0
3	III	30	0,8	1,2	1,1	4,0	3,6
4	IV	90	3,2	1,8	2,0	3,6	4,0

Tables 3.3 and 3.4 present numerical characteristics of fabric porosity and air permeability. The analysis of tables 3.3 and 3.4 shows that the decrease in the number of crossings of threads of one system by another system leads to an increase in the porosity of the fabric and, as a consequence, to an increase in the values of air permeability of clothing fabrics. Similarly, three variants of plain and one variant of finished natural silk fabrics were calculated and their filling parameters are presented in Table 3.5. These samples of natural silk fabrics were produced in the conditions of "Uzbek Scientific Research Institute of Natural Fibres" on the STB type shuttleless weaving machines.

Table 3.5.

Filling parameters of natural silk fabrics

№	Name	Unit.	Indicators of fabric samples			
			I severe	II severe	III severe	IV finished
1	Linear density of the main yarn	tex	2,33x3 twisted Z.S, (2400)	2.33x4 gentle	2.33x4 gentle	2.33x4 gentle
2	Linear density of weft yarn	tex	2,33x3 twisted	2,33x4 twisted Z.S,	2.33x8 twisted Z.S,	2.33x8 twisted

			Z.S, (2400)	(2400)	(2400)	Z.S, (2400)
3	Finished fabric width	see.	65	172	172	170
4	Width of woven fabric	see.	84	176	176	176
5	Filling width across the reed	see.	84	176	176	176
6	Fabric density on the base	n/dm	258	360	360	374
7	Weft density	n/dm	158	330	210	216
8	Number of warp yarns	pcs.	2520	25648	25648	25648
9	Reed number	tooth/dm	150	180	180	180
10	Surface density of the fabric	gr/m^2	51	78	88	73
11	Fabric filling	%	47	56	56	66

The significance of fabric surface density (item 10 of Table 3.5) is that it is a rather sensitive control indicator of the correctness of fabric selection, as it has a significant impact on the hygienic properties of clothes, so it is most rational to use I fabric sample, where the lowest indicator of fabric surface density is used. Technological, physical-mechanical, aesthetic and hygienic researches on the samples of designed and produced fabrics from natural silk have been carried out on the following indicators: yarn processing on the base and on the weft; non wrinkling of fabric; sliding ability of fabric; breaking load of fabric; porosity of fabric; air permeability of fabric. The results of research of natural silk fabrics are presented in Table 3.6. It follows from the table 3.6 that the decrease of fabric density in the sample I allows to reduce the processing of warp threads and as a consequence causes the saving of expensive raw material consumption.

Table 3.6

Characteristics of natural silk fabric samples

Fabric sample options	Indicator values					
	Fabric yield, (%)	Tissue firmness, (%)	Tissue extensibility, (N)	Breaking load of fabric, (N)	Fabric porosity, (%)	Air permeability of the fabric (*cmslms*)
I	4,0/2,0	50/49	17/19	316/474	84	356
II	9,0/1,9	22/23	22/17	464/474	75	221
III	9,0/1,5	22/22	22/18	464/612	77	284
IV	16/4,0	57/57	22/22	404/355	69	156

where: numerator - index for warp; denominator - index for weft.

The value of non-crushability in sample I, due to the twisted main thread, has a significant influence on the preservation of beautiful appearance of fabrics, when compared with other fabric samples. The values of tensile load and mobility of fabrics are within the normative documents for all samples of natural silk fabrics. Also the analysis of tables 3.5 and 3.6 show that fabrics after finishing sample IV in comparison with sample III reduce their air permeability. The greatest influence on the air permeability index is the density of the fabric in warp and weft. Reducing the density of the fabric (sample I) increases its air permeability relative to the other samples II, III, IV. Thus, the results of researches testify to the excellent characteristics of fabric in sample I on such indicators as yarn workmanship of fabric, non-swearing fabric, fabric extensibility, tensile load of fabric, porosity of fabric and air permeability of fabric, which indicators are undoubtedly higher than in other samples or correspond to normative documents. The quality of garment fabrics has also been assessed. Clothing fabrics should possess a complex of physical-mechanical and hygienic properties and wear resistance. During operation they are subjected to repeated mechanical effects (stretching, bending, etc.). In clothing fabrics often use synthetic materials, which worsen their hygienic properties, so the use of natural materials significantly improve these properties, taking into account the climatic conditions of the region. The developed samples of fabrics (10 variants) were tested for research of physical and mechanical properties (stretching, elongation, abrasion) and hygienic properties (air permeability) in the certification centre at TITLP on modern devices according to the developed methodology of laboratory research of fabrics. Table 3.7 shows the results of breaking load on warp (P_{po}) and weft (P_{py}), elongation (l_{po}) and (l_{py}), abrasion and air permeability of fabrics for ten variants (I-X) with variable rapport and number of thread transitions in the weave. In the numerator are presented the performance of fabrics with cotton weft, and in the denominator the performance of fabrics with capron weft.

Table 3.7

Effect of variable rapport and variable thread transition number in a weave on the breathability of a fabric.

№	Weave options	RO	Ry	losr	W	On the basis of		On the duck		Scrubbing, cycle	Air permeability, cm /cm^{32} s
						P_{roo}, thread/ dm	l_{ro}, cm	Rru, yarn /dm	l_{ru}, cm		
1	I	4	4	3,0	3,0	323 328	12,2 11,9	618 1015	13,5 16,6	69 38	54 127
2	II	6	6	4,7	4,7	376 339	12,7 9,7	671 1671	12,9 32,6	74 28	162 260

3	III	6	6	4,7	4,7	363 328	12,7 9,7	654 1634	12,5 32,6	64 22	66 127
4	IV	12	12	9,3	9,3	371 305	12,5 8,7	630 1655	12,5 31,6	61 18	88 156
5	V	12	12	9,3	9,3	371 311	12,5 12,7	626 1634	13,9 31,8	57 19	89 169
6	VI	12	12	7,2	7,2	275 274	10,4 12,0	570 1565	11,9 33,7	35 16	113 237
7	VII	12	12	8,3	9,0	301 305	10,4 8,7	617 1614	12,6 27,8	47 15	123 203
8	VIII	12	12	6,3	5,5	236 269	10,7 14,4	560 1551	10,2 15,9	21 12	134 253
9	IX	12	12	8,8	8,8	320 308	10,1 10,1	599 1604	12,2 22,2	49 17	127 221
10	X	12	24	16	7,0	384 361	13,0 15,2	574 1570	11,8 31,2	53 16	97 182

The analysis of Table 3.7 shows that the physical-mechanical and hygienic properties of the fabric are influenced by the number of warp and weft yarn transitions within the pattern, as well as the type of raw material used in the weft (numerator - cotton, denominator - kapron). The data of breaking load, abrasion and air permeability clearly illustrate that with increasing number of thread transitions ($t_{оср}$ and $t_{уср}$): breaking load on warp and weft increases; abrasion of fabric increases; air permeability of fabric decreases, the exception is variant II, as the weave forms in the fabric like a net.

The use of capron weft leads to an increase in breaking load, elongation and breathability, and to a decrease in abrasion compared to fabrics in which cotton weft is used, due to the physical and mechanical properties of capron thread (breaking load, low coefficient of friction, elongation, etc.). It is reasonable to use samples of II variant as clothing fabrics, as this fabric has good indicators of physical and mechanical properties, hygienic properties and consumer properties. Table 3.8 shows the influence of constant rapport and variable number of thread transition in the weave on the air permeability of the fabric.

Table 3.8

Effect of constant rapport and variable yarn transition number in a weave on the breathability of a fabric.

№	Tissue structure parameters				Air permeability of fabric, cm /cm^{32} s		Surface density of fabric, g/m^2	
	Ro	*Ry*	*to*	*ty*	White Duck	Black Duck	White Duck	Black Duck

1.	8	8	2	2	152	124	147	148
2.	8	8	3	3	130	112	155	156
3.	8	8	4	4	115	103	160	163
4.	8	8	4	4	114	104	158	161
5.	8	8	5	5	103	98	166	168
6.	8	8	6	6	94	77	172	174
7.	8	8	7	7	87	65	179	180

Seven variants of fine-patterned weaves with the rapport $R_o = Ry = 8$, with the number of thread transitions (t_o, t_y) from two to seven (Fig.2.9.), linear density of warp threads $T_o = 20$ tex, fabric density $R_o = 240$ threads/dm, two colours of weft threads - white and black, linear density of weft threads $T_u = 18{,}5x2$ tex and fabric density of weft fabric $R_u = 300$ threads/dm were used. The analysis of Table 3.8 shows that for fine-patterned weaves of fabrics with rapport $R_o = R_y = 8$, when increasing the number of thread transitions (t_o,t_y) from two to seven, in variants with white weft thread the air permeability of the fabric increases by 43%, and in variants with black weft thread the air permeability of the fabric increases by 48%, and the surface density of the fabric remains unchanged. Comparison of white and black weft threads in the fabric shows that the air permeability of the fabric decreases on average by 14%, when using black weft threads in the fabric. The fabric samples developed should meet the aesthetic requirements for fabrics in terms of fashion, colour, material texture, weave and appearance. Teachers, industry specialists, masters, etc. can be used as an expert to assess the aesthetic properties of fabrics. For expert evaluation use the data of survey of **m** - specialists-experts pre-selected n - properties of material variants x_i, **X2**, x_n give a rank estimation of their significance, denote
tea the most important quality indicator with rank **R** = 1, and the least important with rank **R** = **n** (Table 3.9). If any properties, according to the expert's opinion, are equally important, then we take the average of the adjacent ranks and mark each of the properties. The results of the expert survey are entered into a matrix table, which is used to determine the significance of properties and to calculate the coefficient of agreement (concordance), which characterises the consistency of expert assessments.

Table 3.9

Results of the expert evaluation

Experts	About samples of finely patterned fabrics, $n = 7$							
$m = 10$	$X1$	$X2$	Hz	X	$X5$	Hb	$X7$	Σ
1	3	6	2	5	1	7	4	28

2	2	1	4	6	3	7	5	28
3	3	4	2	1	5	6	7	28
4	4	3	1	5	7	2	6	28
5	1	4	3	6	2	7	5	28
6	1	2	3	5	4	6	7	28
7	3	2	4	5	1	6	7	28
8	3	4	2	5	1	6	7	28
9	2	6	7	1	4	5	3	28
10	6	7	4	2	5	3	1	28
Si	28	39	32	41	33	55	52	280
m-n-Si	42	31	38	29	37	15	18	
Yi	0,2	0,148	0,181	0,138	0,176	0,071	0,086	
Yio	0,284	0,209	0,257	-	0,25	-	-	
(S-Si)	12	1	8	-1	7	-15	-12	
$(S-Si)^2$	144	1	64	1	49	225	144	628
Ti								2800

To determine the consistency of expert judgements, the initial data from the matrix table are used. The coefficient of agreement (concordance) is determined by the formula:

$$W = \frac{\sum_{i=1}^{n}(S_i - \bar{S})^2}{\frac{1}{12}\cdot m^2\cdot(n^3 - n) - m\cdot\sum_{j=1}^{m} T_j}$$

Where: $\bar{S}$ - average sum of ranks for all indicators.

We determine the identical assessments of different indicators by individual experts using the formula:

$$\bar{S} = \frac{\sum_{i=1}^{n} S}{n}$$

where: ***u* - *the*** number of ranks with **the** same assessments of the ***i*** -th expert; t_i - the number of assessments with the same rank of the ***i*** -th expert. To assess the significance of the coefficient of agreement (concordance) by Pearson's criterion we determine $\chi_P^2 = 13{,}2 > \chi_T^2 =$ 12.6, we have a significant (meaningful) consistency of the rank estimates of ten experts.

Table 3.10 shows the calculations of evaluation of the quality of fabric samples with constant rapport and constant yarn transition number in the weave on the air permeability of the fabric.

Table 3.10

Effect of constant raport and constant

yarn

transition number

on the air permeability of the fabric.

№	Tissue structure parameters						Air permeability of fabric, cm /cm^{32} s
	Ro Ry	*to ty*	*Po* yarn/dm.	*Ru* yarn/dm	*To Tex*	*Tu tex*	
1.	4	2	258	150	25x2	14,3x5	32
2.	4	2	258	165	25x2	14,3x4	21
3.	4	2	258	190	25x2	14,3x3	17
4.	4	2	258	235	25x2	14,3x2	12
5.	4	2	258	330	25x2	14,3x1	10

Quality comparisons of **m** = 5 fabric samples by the following quality indicators: x1 - fabric appearance; **X2** - surface density of fabric; **X3** - breathability of fabric; x4 - abrasion resistance of fabric; **X5** - workmanship on fabric base; **X6** - workmanship on fabric weft are given in Table 3.11. Each fabric property is evaluated by the rank ***R***, the best property of fabric sample *R*= 1, and the worst property of fabric sample *R*= **m**. Table 3.11 shows the results of natural quality indicators of fabric samples. Of the natural quality indicators of fabric ***n*** = 6, the indicator xi is determined organoleptically, and the rest x2, xz, x4, **X5** and xb - by the instrumental method.

Table 3.11.

Results of quality indicators of fabric samples.

Tissue samples *m* = 5	Natural indicators of fabric quality *n* = 6					
	X1-	*X2*	*Hz*	*X4*	*X5*	X6
	Fabric appearance	Surface density of fabric gr/m^2	Air permeability of fabric, cm /cm^{32} s	Tissue abrasion, cycle	Yield by base, %	Weft workmanship %
1	2	270	32	6900	9,5	2,9
2	3	260	21	7300	8,7	3,9
3	1	246	17	10700	8,4	4,4
4	4	213	12	14000	6,0	6,1
5	5	191	10	16000	4,8	5,7

Table 3.12.

Results of quality assessment of fabric samples.

Tissue samples *m* = 5	Ranking scores of quality indicators of fabric samples *R* 6						ΣR	Location
	XI	*X2*	*Hz*	*X4*	*X5*	*Hb*		

							1	
1	2	5	1	5	5	1	19	5
2	1	3	3	3	3	3	16	1
3	3	4	2	4	4	2	19	4
4	4	2	4	2	2	4	18	2
5	5	1	5	1	1	5	18	3
$\sum_{1}^{5} R$	15	15	15	15	15	15	90	-

Table 3.12 shows the results of quality assessment of fabric samples. From where it follows that the compared fabric samples in terms of quality in the direction of its deterioration are arranged in the following order: 2-4-5-3-1. We specify the quality assessment using the significance coefficients of individual indicators of fabric samples, which are presented in Table 3.13.

Table 3.13

Significance coefficients of individual indicators of fabric samples.

Tissue samples $m = 5$	Ranking scores of quality indicators Ry						Σ ***RY***	Location
	X1	*X2*	*Hz*	*X4*	*X5*	*Hb*		
1	0,4	0,5	0,2	1,0	0,75	0,15	3,1	4
2	0,2	0,3	0,6	0,6	0,45	0,45	2,6	1
3	0,6	0,4	0,4	0,8	0,6	0,3	3,1	3
4	0,8	0,2	0,8	0,4	0,3	0,6	3,1	2
5	1,0	0,1	1,0	0,2	0,15	0,75	3,2	5
Γ	0,2	0,1	0,2	0,2	0,15	0,15	-	-

In terms of quality in the direction of deterioration, the fabric samples (according to Table 3.13) are arranged in the following order: 2-4-3-1-5. The evaluation of the two best samples of fabrics remained unchanged, and the best is sample 2, with the following parameters: surface density of fabric246 gr/m^2 ; air permeability of fabric17 cm /cm^{32} s; abrasion resistance of fabric10700 cycle; processing on the base of fabric 8.4%; processing on the weft of fabric4.4%; porosity of fabric 66%.

In the fourth chapter the warp tension is investigated and optimisation of the process of forming garment fabrics is carried out. One of the main technological parameters influencing the technological process and fabric structure is the warp tension as the winding winding is triggered on the warp, on the filling width and per machine cycle. As insignificant deviation from the set value of these parameters leads to increase of thread breakage and decrease of quality of produced fabrics. Therefore, it is advisable to carry out studies of warp tension across the filling width of the weaving machine. The tension of the individual threads is different in its magnitude and can be higher or lower than the required tension at any section of the threading width, i.e. there are weakly and strongly tensioned threads along with normally tensioned threads. During the study, a line parallel to the axis of rotation of the warp was drawn on the winding of the weaving warp with paint and brought to the earning of the fabric. We had to work the fabric at different filling tension. The filling tension was changed by means of notches on the lever of the weaving machine's main regulator. The number of notches varied from two (2) to six (6). After the line drawn on the winding slip was worked into the fabric, the filling tension was changed, the line was again drawn on the winding slip and worked into the fabric. In this way, five pieces of fabric were worked at five different warp tensions. As the threads progressed along the filling line, the marks shifted as the highly tensioned threads lagged behind and the weakly tensioned threads moved forward. This resulted in a band of mixed marks in the fabric on the line drawn on the winding line of the weaving warp. The instruments of the study were a ruler, a magnifying glass and a needle. The study was carried out in this way: in the middle of the cloth, in the same place (since we have five webs), a certain number of main threads were selected. Specifically, one hundred and ninety-eight (198 pieces) main threads were investigated, as it was not practical to investigate all threads. After that, a single weft was selected along the middle of the strip in the direction of the weft yarns as the line against which the mark bias was determined. The amount of mark displacement was determined by the number of wefts by which the mark on the main thread was displaced. From the measurement results it follows that the marks on the main lines are scattered to one side or the other along the centre line by different numbers of thins along the length of the fabric. This shows that the individual warp threads have different tensions, i.e. the warp threads whose marks remained up to the centre line along the length of the fabric are strongly taut, and those threads whose marks moved from the centre line along the length of the fabric towards the sternum are weakly taut. The main threads whose marks were on the centre line are considered to be normally taut. But in practice such high accuracy is impossible, and in this case such accuracy is not reasonable. Therefore, we took as normal tension those yarns whose marks were placed within four points to one side or the other of the centre line along the length of the fabric. Table 4.1 shows the results of counting the number of differently tensioned warp yarns in the fabric, which shows that the warp yarn tension across the width of the fabric is made up of the tension of medium tensioned (normal tensioned), high and low tensioned yarns, this does not contradict our assumptions that the size of the filling tension has a significant effect on the unevenness of warp yarn tension. Most warp yarns are medium tension yarns. For example, according to the calculation at the fourth (4) notch, 59 % of the yarns are normally tensioned, and 21 % and 20 % of the yarns are highly tensioned and slightly tensioned, respectively (see Table 4.1).

Table 4.1.

Number of differently stretched warp yarns in the fabric

	Number of notches, values.									
	2		3				5		6	
Warp tension	abs.	%	abs.	%	abs.	%	abs.	%	abs.	%
Tight strands	50	25	45	23	41	21	36	18	33	17
Medium tension threads	89	45	103	52	117	59	127	64	133	67
Loose threads	59	30	50	25	40	20	35	18	32	16
Total	198	100	198	100	198	100	198	100	198	100

Using the data from Table 4.1, we made a graph of the dependence of the number of moderately tensioned, strongly tensioned and weakly tensioned threads on the value of threading tension (Fig.4.1 - 4.3). On the abscissa axis, the threading tension (number of notches) was plotted, and on the ordinate axis, the number of threads (as a percentage of the total number of threads examined). As the warp tension increases, the percentage of strongly and weakly tensioned threads across the width of the fabric decreases, while the percentage of moderately tensioned threads across the width of the fabric increases.

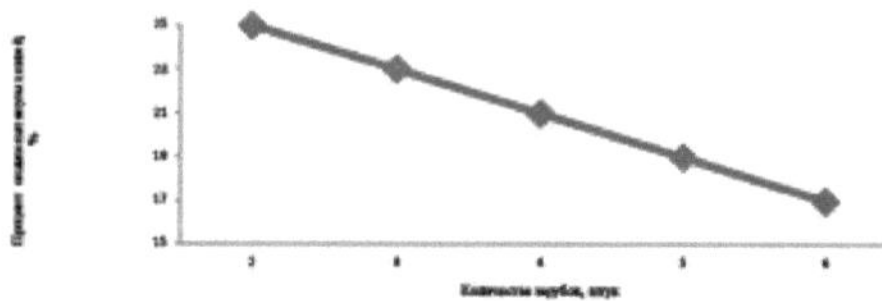

Fig.4.1 Graph of the effect of the number of notches (filling tension) on the percentage of heavily tensioned threads across the width of the fabric

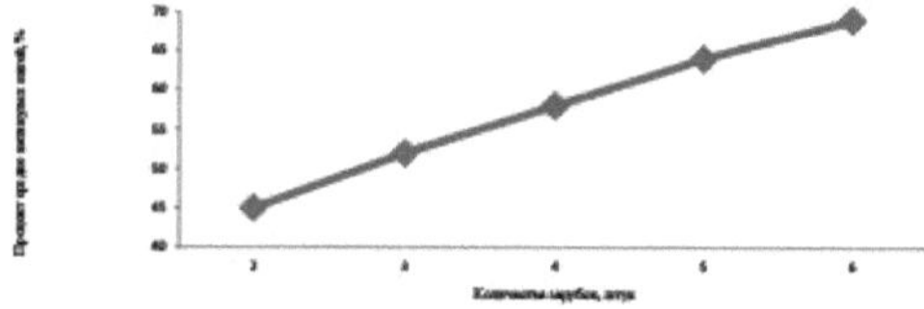

Fig.4.1 Graph of the effect of the number of notches (filling tension) on the

percentage of heavily tensioned threads across the width of the fabric

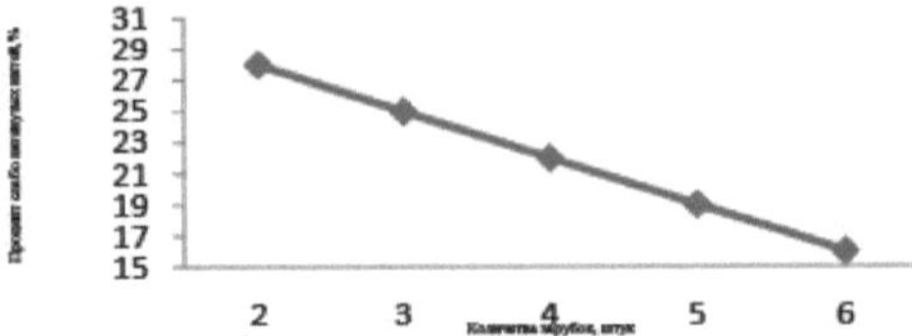

Fig.4.3 Graph of the effect of the number of notches (filling tension) on the percentage of loosely stretched threads across the width of the fabric

When determining the regression model for the object with one input and one output, we conducted an active experiment in a wide range of changes in the factor X. The number of levels of the factor or the number of experiments in the experiment planning matrix is taken N = 5. To improve the accuracy of determining the output parameter Y, each experiment of the matrix is repeated several times, m = 5. Let us consider the operation in which we studied Y - absolute values of the number of average tensioned warp threads in the fabric depending on the filling tension of warp threads (number of nicks of the main regulator lever) X on the weaving machine (see table 4.1). As a result of the mathematical analysis, equations describing the relationship between the warp filling tension and the number of high, medium and low warp tension yarns were obtained.

For normally tensioned yarns YR = 69 + 11.2-X (4.1)

For highly stretched yarns: YR = 58.2 - 4.3 -X (4.2)

For weakly stretched yarns: YR = 70.8 - 6.9 -X (4.3)

What follows from these equations is that the filling tension of the warp and the irregularity of the single warp yarns are related by a straight-line relationship. Figs. 4.4 - 4.6 show the graphical dependence of the number of medium tension, high tension and low tension yarns across the width of the fabric on the filling tension of the warp yarns. The graphs clearly illustrate that with increasing filling tension the number of medium tensioned threads increases, while the number of high and low tensioned threads decreases.

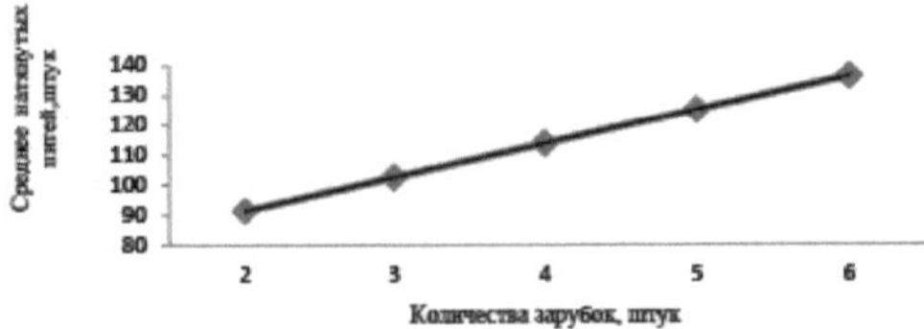

Fig. 4.4. Dependence of the number of average tensioned threads on the filling tension

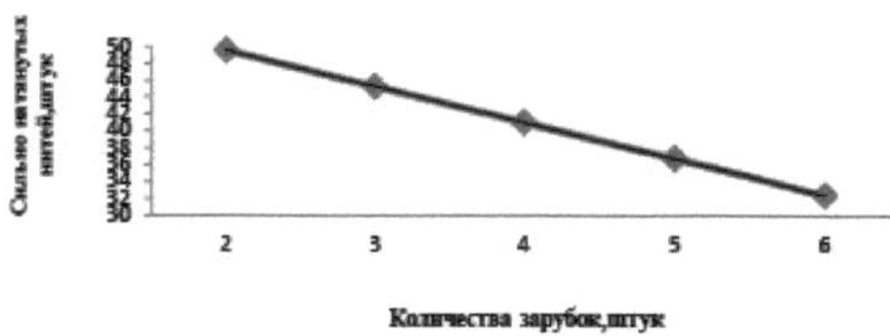

Figure 4.5. Dependence of the number of heavily tensioned threads on the filling threads

tensions

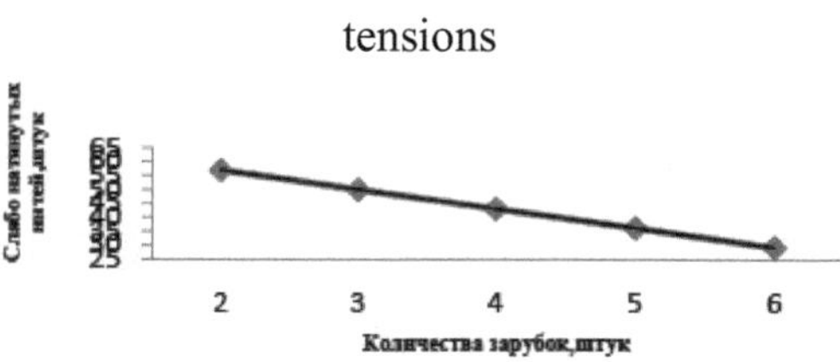

Fig.4.6. Dependence of the number of loosely tensioned threads on the filling tension

The practice of using devices based on the mechanical principle of action shows that this group of devices has rather low sensitivity, significant inertia of moving parts and does not provide registration of tension changes of threads with a small oscillation period. Devices of the mechanical principle of action can be used for approximate measurements of tension of threads with a large oscillation period. Mechano-optical instruments for measuring yarn tension have become very common in warp tension studies on the weaving machine. These instruments compare favourably with mechanically operated instruments in that the mass (weight) of the moving parts is reduced to a minimum. Mechanical-optical devices quite satisfactorily catch the warp tension fluctuations in the weaving process and allow to observe the changing tension value both visually and to register the tension fluctuations on photographic paper or photographic film. We have developed a tool for measuring the tension of single threads of the mechanical principle of action on the basis of the Uster device (Fig. 4.7), the distinctive feature of which is the presence of a platform 1, where two devices 2 are installed to measure the small and large value of tension of threads 3, and force-measuring thread guides are made in the form of one-shoulder levers 4. The advantage of the new means is the accuracy of measuring the tension of single threads to obtain quantitative characteristics, simplicity and ease of maintenance.

Fig. 4.7: Instrument for measuring the tension of single main threads on the weaving machine.

In this work, warp tension measurements were made with a new single thread tension gauge in zones evenly distributed across the loom filling width (Fig. 4.8).

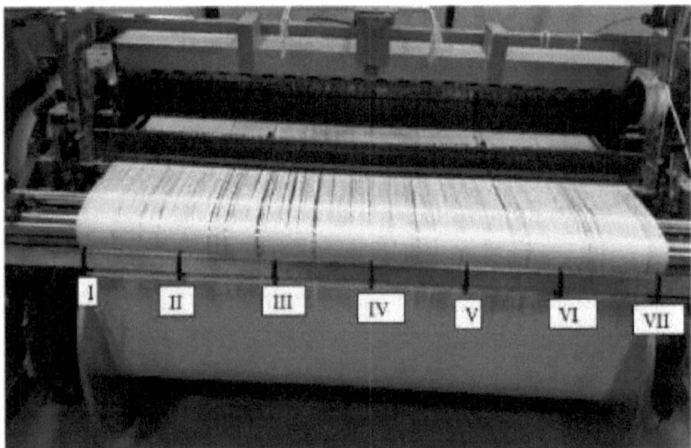

Fig. 4.8 Distribution of the yarn tension measuring zones across the weaving machine insertion width

Table 4.8 shows the distribution of yarn tension measurement zones over the filling width of the machine. Average warp tension values were obtained by zone and per machine cycle. These measurements and calculations were performed at least five times. Then we determined the average value of yarn tension and the variance of the average value of yarn tension. The error of the obtained values was within 5%.

Table 4.2.

Distribution of the yarn tension measurement zones across the filling width of the machine.

	Numbering of the thread	Distances of the yarn tension measuring zones across the

	tension measuring zones across the filling width of the weaving machine	filling width of the machine on the filling side (left to right), cm.			
		Weaving machine type			
		AT-100-5M	ATPR-100	STB-175.	P-190
1	I	5	5	5	5
2	II	20	20	30	30
3	III	35	35	55	55
4	IV	50	50	80	80
5	V	65	65	105	105
6	VI	80	80	130	130
7	VII	95	95	155	155

Tables 4.3-4.6 show the average tension values across the filling width and their dispersion for fabrics produced on different types of weaving machines. According to these data, the warp tension has a large unevenness across the filling width. The unevenness of tension is influenced by the method of weft insertion and the type of shedding mechanism.

Changing the warp tension over the threading width for shuttle machines

Thread tension measurement zones across the filling width of the weaving machine	Indicators of warp tension values across the width of the fabric					
	average value U, *cH*			Dispersion of the mean S value^2 ^), *cH*		
	In the surf.	When yawning.	With a seizure.	In the surf.	When yawning.	With a seizure.
I	69,8	39,6	16,7	8,5	4,8	2,0
II	24,3	21,8	16,8	3,00	0,7	2,0
III	39,7	36,2	14,2	4,8	4,4	1,7
IV	28,1	25,0	12,2	3,4	3,0	2,4
V	37,9	33,1	14,0	4,6	4,0	1,7
VI	30,3	26,2	18,9	3,7	3,2	2,3
VII	75,3	36,2	20,2	9,2	4,4	2,4

Table 4.4.

Variation of warp tension over the filling width for air-jet machines

Thread tension measurement zones across the filling width of the weaving machine		Indicators of warp tension values across the width of the fabric					
		Average U value, *cH*			Dispersion of the mean value of S^2 ^), *cH*		
		In the surf.	When yawning.	With a seizure.	In the surf.	When yawning.	With a seizure.
1	I	30,7	28,6	22,8	3,7	3,6	2,7
2	II	33,6	30,2	23,2	4,1	3,8	2,8
3	III	40,0	23,7	27,4	4,9	4,2	3,3
4	IV	49,0	43,2	30,2	6,0	5,4	3,7

5	V	39,2	35,6	24,1	4,8	4,4	2,9
6	VI	34,0	30,0	23,4	4,1	3,7	2,8
7	VII	32,1	29,0	22,5	3,9	3,6	2,7

Table 4.5.

Variation of warp tension over the filling width for microstitch machines

Thread tension measurement zones across the filling width of the weaving machine		Indicators of warp tension values across the width of the fabric					
		Average U value, *cH*			Dispersion of the mean value of S^2 ^), *cH*		
		In the surf.	When yawning.	With a seizure.	In the surf.	When yawning.	With a seizure.
1	I	33,3	30,1	25,6	3,2	3,0	2,7
2	II	41,7	38,2	27,1	3,7	3,3	3,0
3	III	44,1	41,1	31,2	3,6	3,7	3,1
4	IV	48,6	44,7	34,1	4,0	4,0	3,5
5	V	40,6	36,9	29,9	3,9	3,7	3,1
6	VI	36,8	32,4	27,8	3,7	3,3	3,1
7	VII	30,2	29,1	26,2	2,8	2,8	2,7

Variation of warp tension across the filling width for rapier machines

Thread tension measurement zones across the filling width of the weaving machine		Indicators of warp tension values across the width of the fabric					
		Average U value, *cH*			Dispersion of mean value of S^2 (V), *cH*		
		In the surf.	When yawning.	With a seizure.	In the surf.	When yawning.	With a seizure.
1	I	26,5	32,4	27,2	3,5	3,5	3,0
2	II	44,5	40,1	29,8	4,1	3,8	3,2
3	III	48,3	43,8	29,8	3,8	3,6	3,2
4	IV	49,8	46,2	33,3	4,2	4,1	3,5
5	V	46,1	41,4	30,1	4,1	4,0	3,7
6	VI	41,8	38,7	30,1	3,9	3,9	2,9
7	VII	37,6	33,9	28,3	3,3	3,2	3,0

Regression models of the dependence of the thread tension of the base *yarnsU* on the filling width *x* at the moment of surfing were also obtained:

For shuttle machines of AT type $U = 67,8 - 2,14 \cdot x + 0,024 \cdot x^2$ (4.4)

For pneumatic sawing machines of ATPR type $U = 30,4 + 0,76 \cdot \chi - 0,008 \cdot x^2$ (4.5) For microstitching machines of STB type $U = 34,3 + 0,41 \cdot \chi - 0,003$ - (4.6)

For rapier machines of type P $U = 36,2 + 0,34 \cdot \chi - 0,002 \cdot \chi^2$ (4.7)

An analysis of the graphs (Fig. 4.9-4.12) of the equations shows that the change in warp tension across the filling width is parabolic, hence the warp tension is not uniform. On shuttleless machines (air-jet, micro-jet and rapier), the minimum warp tension is at the fabric edges, and the maximum tension is in the middle of the threading. On shuttle machines, on the contrary, the warp tension is maximum at the fabric edges and minimum in the middle of the threading.

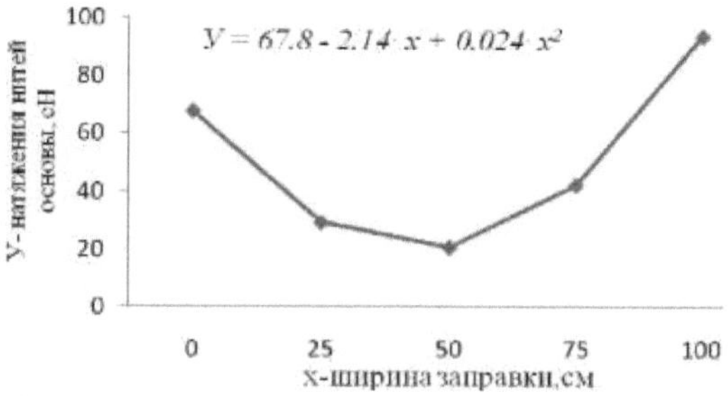

Fig. 4.9. Regularity of warp tension variation along the filling width of shuttle machines

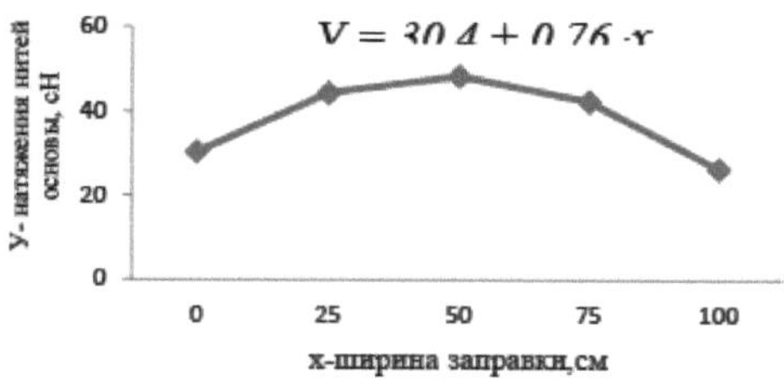

Fig.4.10. Regularity of change of warp tension along the filling width of air-jet sawing machines

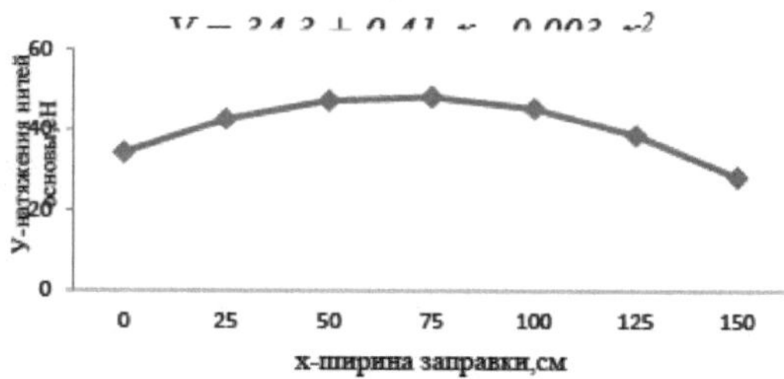

Fig. 4.11. Regularity of warp tension variation across the filling width of microstitch machines

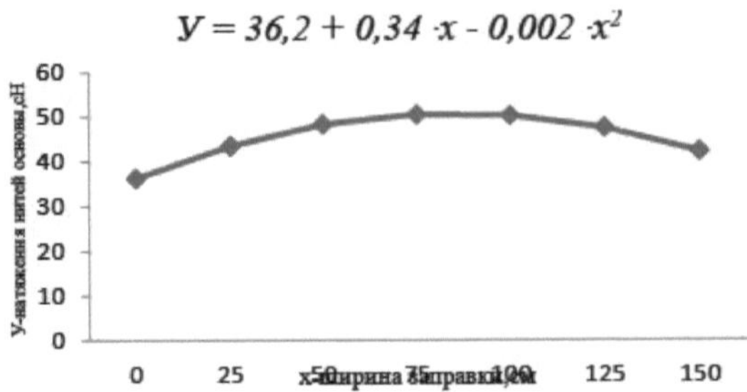

Fig. 4.12. Regularity of warp tension variation across the width of the rapier loom dressings

Experimental studies on the filling width of the weaving machine have shown that the tension is not uniform across the filling width. It depends on the conditions of warp preparation for weaving and the different physical and mechanical properties of the yarns. In addition, the warp yarns picked into the looms have a different movement at the height of each loom. Existing mechanisms for levelling the yarn tension across the width of the threading are inefficient and therefore have not found application in the industry. In order to improve the uniformity of the warp tension across the filling width, we propose a means for levelling the warp tension across the filling width of the loom in the form of an elastic material in contact with the warp threads. The most rational fitting of the proposed device on the loom is carried out by a scalo. The essence of the developed construction is explained by the drawing Fig.4.13. The warp threads 1 envelop the elastic material 2 in the cavity of which the scalo 3 is located. The levelling ability of the mechanism is determined by the elasticity and thickness of the material, which is selected depending on the range of warp yarns to be processed. During weaving machine operation, the warp tension is absorbed by the elastic material 2.

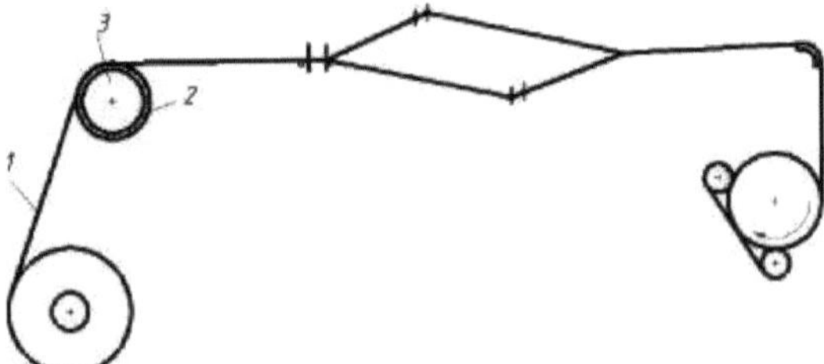

Fig. 4.13. Warp tension stabiliser across the filling width of the weaving machines

At the same time, the effect of individual threads on the latter will be different depending on the tension. In places of high thread tension the elastic material deforms, absorbing the tension changes, and in places of the lowest tension due to the elasticity of the elastic material it increases to the set value. Consequently, the peak values of tension decrease and the minimum values increase to the average for all values, i.e. there is an equalisation of the individual warp threads. As a result, the evenness of warp tension is improved, warp breakage is reduced and the quality of the fabric produced is improved. The different warp tension across the filling width of the machine makes it somewhat difficult to study this problem, as there are a large number of warp threads in the warp with a wide range of tension changes. Therefore, we will make some assumptions in the analytical study. As it is known from physics of high-molecular compounds, the most convenient methods of studying tensions and character of their superposition in textile materials are analogues in the form of mechanical models. There are a variety of models that roughly demonstrate the nature of tension changes in textile materials. More suitable for practical application, the visco-elastic model of Dogadkin B.A., in which the elastic element (spring) is connected in parallel with the Maxwell model (serial connection of the spring with the shock absorber). The parallel connection of the elements in the model allows us to write down the following condition, in case of increasing tension of a single warp thread of the elastic filling system: $T = T_3 + \Delta T_y - \Delta T_p$ (4.8), where T is the current tension of the elastic filling system; $T_з$ is the filling tension of the elastic filling system; ΔT_u is a random change in the tension of the elastic filling system; ΔT_r is the change in tension from the elastic element to the scalo.

The warp tension depends on the deformation and stiffness of the elastic filling system and has the following dependencies

$$T = \lambda \cdot C \qquad (4.9)$$

Substituting the obtained expression, we obtain

$$T_{max} = \lambda_з \cdot C_з + \Delta\lambda_y \cdot C_y - \Delta\lambda_p \cdot C_p \qquad (4.10)$$

where: $\Delta\lambda_p$ - elastic deformation of the element on the scalo; λ_u - warp yarn deformation; C_y - stiffness coefficient of the warp yarns; C_p - stiffness coefficient of the elastic element.

In the case of decreasing tension of a single thread in an elastic threading system, we obtain the following equation $T_{min} = \lambda_з \cdot C_з - \Delta\lambda_y \cdot C_y + \Delta\lambda_p \cdot C_p$ (4.11)

Analysing the obtained equation we can say that at random increase or decrease of tension of separate threads along the width of the machine dressing the elastic material of tension elastic element, on the scalo, deforming, compensates the length of warp in the elastic system of dressing, at decrease of tension of threads, straightening, elastic element, restores the tension of this thread to the average value. The unevenness of the warp tension is determined by

$$\delta = \frac{T_{max} - T_{min}}{T_з} \cdot 100\% \qquad (4.12)$$

or after substitution and transformation we have

$$\delta = \frac{\Delta\lambda_y \cdot C_y - \Delta\lambda_p \cdot C_p}{\lambda_з \cdot C_з} \cdot 100\% \qquad (4.13)$$

Let us analyse the obtained equation. In existing designs, the scalo is absolutely rigid and the deformation of the scalo is zero. Consequently, there is a large unevenness in the tension of the individual threads across the weaving machine threading width. In the modernised design, the scalo has an elastic element, hence the sensitivity is increased with the variation of the individual thread tension across the loom width. The efficiency of sensitivity depends on the stiffness coefficient of the elastic material, the higher the stiffness, the lower the sensitivity. Here is an example of calculation of deformation of material (rubber-like rubber) with modulus of elasticity $E = 16{,}8 \frac{гр}{мм^2}$, material thickness S = *10mm*, when producing fabric with linear density of the main yarns T_o = *30tex* .

Material deformation $\Delta\lambda_p = \frac{S \cdot T}{F \cdot E}$ (4.14)

$$F = d_o \cdot l_o, мм^2 \qquad (4.15)$$

where: E - modulus of elasticity of the material; T - tension of the main threads; *5* - thickness of the elastic material; F - contact area of the thread with the elastic material;

d_o - thread diameter; l_o - thread length.

$$l_o = \pi d_{CK} \cdot 0{,}25 = 3{,}14 \cdot 150 \cdot 0{,}25 = 117{,}75$$

$$d_o = 0.0316 \cdot \sqrt{T_o} \cdot 1.25 = 0.0395 \cdot \sqrt{T_o}$$

where d_{CK} = *150 mm*

After substitution, we have $F = 0.0395 \cdot \sqrt{T_o} \cdot 117.75 = 4.65 \cdot \sqrt{T_o}$

$$\Delta\lambda_p = \frac{10 \cdot T}{4.65 \cdot \sqrt{T_o} \cdot 16.8} = 0.128 \frac{T}{\sqrt{T_o}} \qquad (4.16)$$

Substituting in (4.14), we have

At T = *const* we calculate $\Delta\lambda_p$ at different linear densities of the main yarns. Table 4.7 shows calculations of the deformation of the elastic material (rubber-like rubber) depending on the linear density of the cotton yarn at a filling tension of the warp yarns T = 20 *cN*.

Table 4.7

Elastic material deformations as a function of yarn linear density.

1	Linear yarn density T_o, *tex*	20	30	40	50	60	70
2	Elastic material deformation $\Delta\lambda_p$, *mm*	0,57	0,47	0,41	0,36	0,33	0,31

Analysis of the obtained data showed that when using the modernised scalo design, the unevenness of tension is sharply reduced. Table 4.8- 4.11 shows the average values of tension across the filling width and their dispersions.

Table 4.8.

Changing the warp tension across the filling width for shuttle looms

Thread tension measurement zones across the filling width of the weaving machine		Indicators of warp tension values across the width of the fabric					
		Average U value, *cH*			Dispersion of the mean value of S^2 ^), *cH*		
		In the surf.	When yawning.	With a stiffness.	In the surf.	When yawning.	With a seizure.
1	I	31,6	29,7	20,9	4,6	5,4	3,17
2	II	27,1	26,5	19,3	3,9	4,9	2,86
3	III	26,1	26,3	18,6	3,3	4,3	2,6
4	IV	24,3	22,7	17,3	2,8	3,8	2,3
5	V	25,9	23,8	18,0	3,0	4,4	2,6
6	VI	29,7	27,0	20,3	3,4	5,1	3,1
7	VII	32,9	29,9	22,1	4,4	5,7	3,3

Table 4.9

Changing the warp tension across the filling width for air saws

Thread tension measurement zones across the filling width of the weaving machine		Indicators of warp tension values across the width of the fabric					
		Average value U, *Cn*			Dispersion of the mean value of S^2 ^), *cH*		
		In the surf.	When yawning.	With a seizure.	In the surf.	When yawning.	With a seizure.
1	I	51,4	43,2	26,4	6,1	5,2	3,2

2	II	51,6	44,0	27,0	6,3	5,3	3,3
3	III	49,2	45,2	28,3	6,0	5,5	3,4
4	IV	51,6	45,9	28,8	6,3	5,6	3,5
5	V	49,2	44,7	28,5	6,0	5,4	3,4
6	VI	51,6	45,0	27,8	6,3	5,5	3,4
7	VII	50,7	44,1	26,9	6,2	5,4	3,2

Table 4.10.

Changing the warp tension across the filling width for microstitching machines

Thread tension measurement zones across the filling width of the weaving machine		Indicators of warp tension values across the width of the fabric					
		Average U value, *cH*			Dispersion of the mean value of S^2 ^), *cH*		
		In the surf.	When yawning.	With a seizure.	In the surf.	When yawning.	With a stiffness.
1	I	34,7	30,9	25,6	3,7	2,9	2,7
2	II	37,3	35,1	26,9	3,6	3,3	2,7
3	III	39,6	37,2	29,1	3,9	3,4	2,9
4	IV	38,2	34,8	28,6	3,5	3,5	3,1
5	V	36,9	33,1	27,2	3,6	3,1	2,9
6	VI	33,5	31,1	25,8	3,3	3,2	2,8
7	VII	32,1	29,3	24,5	3,2	3,1	2,6

Variation of warp tension across the filling width for rapier machines

Thread tension measuring zones across the filling width of the weaving machine		Indicators of warp tension values across the width of the fabric					
		Average U value, *cH*			Dispersion of mean value of S^2 (V), *cH*		
		In the surf.	When yawning.	With a seizure.	In the surf.	When yawning.	With a seizure.
1	I	37,2	33,9	27,3	3,9	3,1	2,5
2	II	39,2	36,1	28,8	4,1	3,3	2,7
3	III	40,7	37,7	29,6	4,1	3,4	2,8
4	IV	40,6	37,2	29,1	4,3	3,7	3,0
5	V	38,4	34,2	27,7	4,1	3,4	3,0
6	VI	37,1	33,8	27,5	3,9	3,3	2,7
7	VII	37,0	33,5	27,4	3,6	3,2	2,6

Regression models of the dependence of warp tension on filling width at the moment of surfing were obtained.

For shuttle machines of AT type $U = 30,3 - 0,29 \cdot \chi + 0,0036 \cdot \chi^2$ (4.17)

For pneumatic sawing machines of ATPR type $U = 32,7 + 0,15 \cdot x - 0,0016 \cdot x^2$ (4.18)

For microstitching machines of STB type $U = 34,3 + 0,13 \cdot x - 0,001 \cdot \chi^2$ (4.19)

For rapier machines of R-190 type $U = 37{,}8 + 0{,}1 \cdot \chi - 0{,}0007 \cdot \chi^2$ (4.20)

Analysis of the obtained models shows that the tension variation along the width of the filling has a curvilinear character, but the unevenness of tension is much lower (Fig.4.14-4.17) than with the existing scalo design.

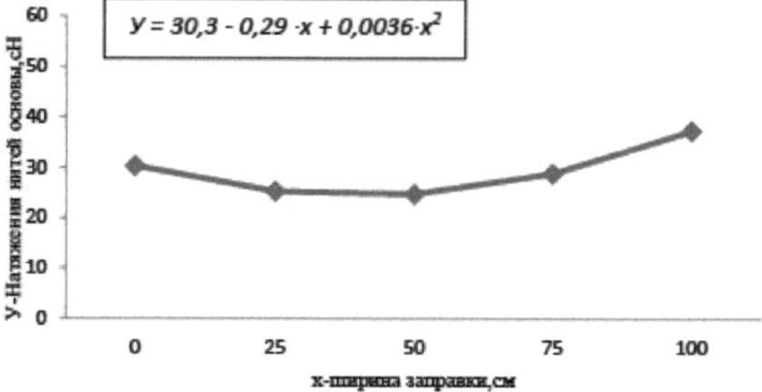

Fig.4.14 Regularity of change of tension of warp threads on width of threading of shuttle machines of AT type after modernisation

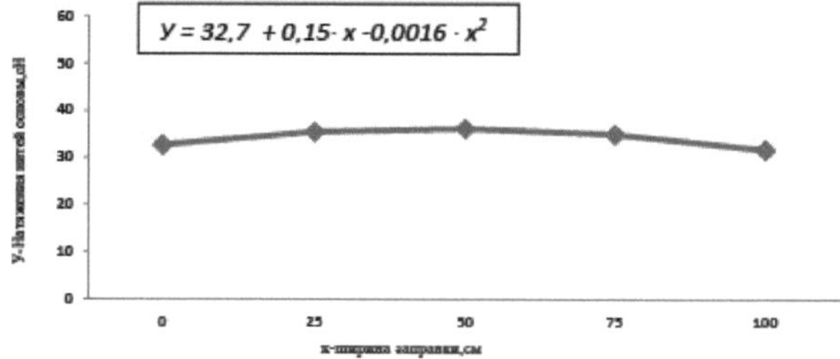

Fig.4.15 Regularity of change of warp tension along the filling width of ATPR type air-jet sawing machines after modernisation

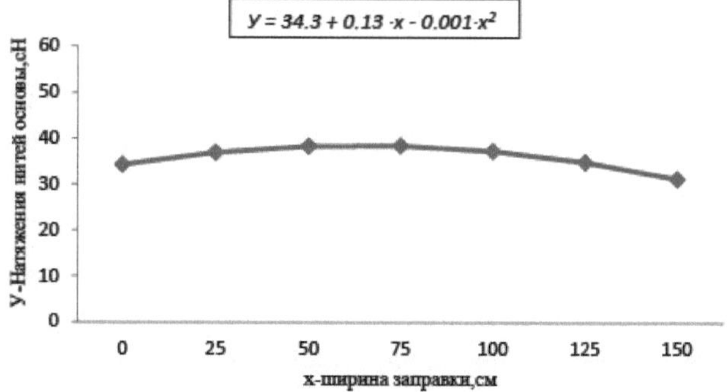

Fig.4.16. Regularity of change of warp tension along the threading width of STB type microstitching machines after modernisation

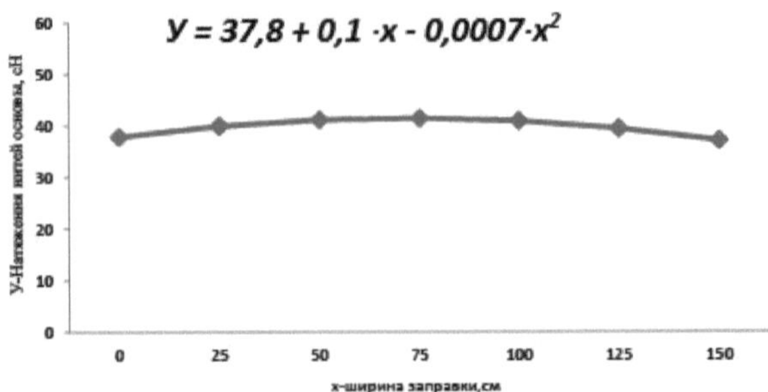

Fig.4.17. Regularity of change of warp tension along the filling width of rapier looms of P type after modernisation

The analysis of the conducted research shows that the uniformity of warp tension across the filling width in the modernised scalo system is significantly higher than in the existing scalo system. A second order central composite method of experiment planning has been adopted in the work, which enables a detailed study, description and optimisation of the weaving process in the investigated area. weaving process in the optimisation domain under study. The selection of intervals and values of factors for five levels of variation was carried out from the consideration of technological possibilities of loom fuelling (Table 4.12)

Table 4.12.

Levels of variation of factors

Factors	Levels of variation					InterVal
	-1,682	-1,0	0	+1,0	+1,682	
x_1 - filling tension of the warp, cN	13	16	20	24	27	4
x_2 - offset value, mm	7	10	15	20	23	5
x_3 - position of the rock relative to the sternum, mm	-15	-10	0	+10	+25	15

The experiment allows us to obtain a second-order mathematical model describing the influence of factors x_1, x_2, x3 on the selected optimisation parameters. An adequate mathematical model describing the dependence of breakage on the selected significant factors is obtained

$$y_R = 0{,}2 + 0{,}03X_1 - 0{,}01X_2 - 0{,}01X_3 - 0{,}02X_1 \cdot X_2 + \\ + 0{,}02X_1 \cdot X_3 + 0{,}03X_2 \cdot X_3 + 0{,}08X_1^2 + 0{,}02X_2^2 - 0{,}02X_3^2 \qquad (4.21)$$

The evaluation of the process experiment was carried out by means of cuts: $U = f(x_1)$ at constant x_2, x_3; $U = f(x_2)$ at constantx_1,x_3; $U = f(x3)$ at constant x_1,

x_2. In Figs. 4.18-4.20 are plotted using formula (4.21).

Turnover per 1 metre of fabric

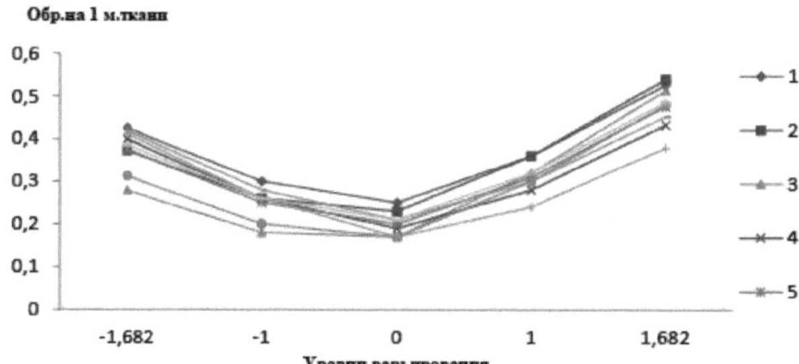

Ряд 1 – X_2 = -1; X_3 = -1. Ряд 2 - X_2 = -1; X_3 = 0. Ряд 3 - X_2 = -1; X_3 = 1.
Ряд 4 - X_2 = 0; X_3 = -1. Ряд 5 - X_2 = 0; X_3 = 0. Ряд 6 - X_2 = 0; X_3 = 1.
Ряд 7 - X_2 = 1; X_3 = -1. Ряд 8 - X_2 = 1; X_3 = 0. Ряд 9 - X_2 = 1; X_3 = 1.

Figure 4.18: Effect of warp filling tension on thread breakage.

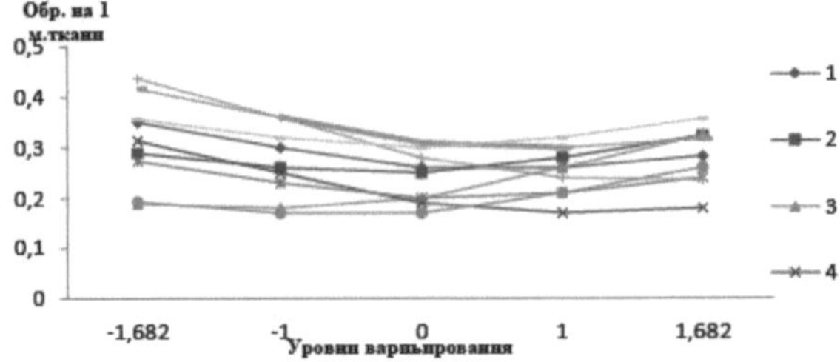

Ряд 1 - $X_1 = -1$; $X_3 = -1$. Ряд 2 - $X_1 = -1$; $X_3 = 0$. Ряд 3 - $X_1 = -1$; $X_3 = 1$.
Ряд 4 - $X_1 = 0$; $X_3 = -1$. Ряд 5 - $X_1 = 0$; $X_3 = 0$. Ряд 6 - $X_1 = 0$; $X_3 = 1$.
Ряд 7 - $X_1 = 1$; $X_3 = -1$. Ряд 8 - $X_1 = 1$; $X_3 = 0$. Ряд 9 - $X_1 = 1$; $X_3 = 1$.

Fig. 4.19. Influence of the size of the gap on thread breakage

The influence of x_3 (position of the scalp relative to the breast of the weaving machine Fig. 4.20.) on y is represented by a convex parabola with minimum values of Y at x_3 equal to - 1.682 and + 1.682 respectively. Consequently, when producing this fabric, it is advisable to raise the scala as much as possible, or lower it as much as possible with respect to the sternum, and raising the scala ($x_3 = +1.682$), as shown in Fig. 4.20 results in the lowest breakage. The curve of change of breakage y from the position of the offset x2 (Fig. 4.19) at zero value x_1 (the tension of the main threads) and the maximum raised rock $x_3 = +1{,}682$ shows that at x2 = -1 it is possible to reduce the breakage of threads in 2 times, ie the parameters will have the following values: the tension of the main threads - 16 cN (per 1 thread); the value of the offset - 10 mm.; the position of the rock relative to the sternum - (+ 25) mm.

Obr.per 1 metre of fabric 0.4

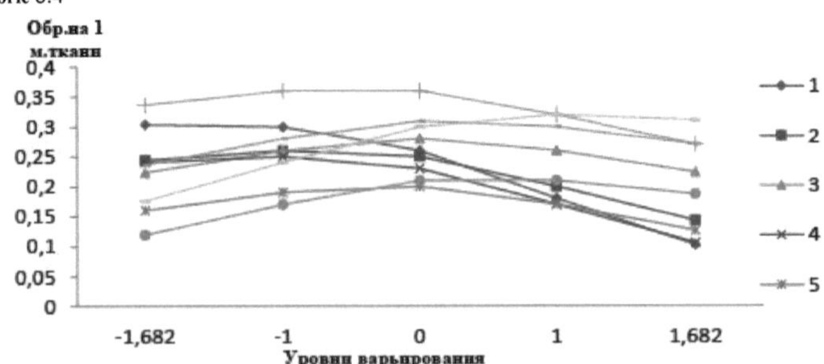

Ряд 1 - $X_1 = -1$; $X_2 = -1$. Ряд 2 - $X_1 = -1$; $X_2 = 0$. Ряд 3 - $X_1 = -1$; $X_2 = 1$.
Ряд 4 - $X_1 = 0$; $X_2 = -1$. Ряд 5 - $X_1 = 0$; $X_2 = 0$. Ряд 6 - $X_1 = 0$; $X_2 = 1$.
Ряд 7 - $X_1 = 1$; $X_2 = -1$. Ряд 8 - $X_1 = 1$; $X_2 = 0$. Ряд 9 - $X_1 = 1$; $X_2 = 1$.

Row 1 - X1 = -1; X2 = -1. Row 2 - x_1 =-1;X2= 0.Row 3 - X1 =-1; X2=1. =-1;x_2= 0.Row 3 - X1 =-1;X2=1.

Row 4 - $X1 = 0$; $X2 = -1$. Row 5 - $x1 = 0; x2 = 0$ .Row 6 - $X1 = 0; X2=1$

.

Row 7 - $X1 = 1$; $X2 = -1$. Row 8 - $x1 =1; X2= 0$.Row 9 - $X1 = 1$; $X2=1$.
$=1; x2= 0$. Row 9 - $X1 = 1; X2=1$.

Fig.4.20.Influence of the scalo position on thread breakage

With these parameter values, the breakage of the main threads will not exceed 0.1 breaks per 1 metre of fabric.

CONCLUSION

The analysis of literary sources shows that mainly the works are directed to the research of structure and designing of fabrics according to the given thickness, fabric filling, strength, order of phases of fabric structure, linear density of threads, degree of unbalance. The structure and design of garment fabrics have not been sufficiently investigated, so it is advisable to carry out the design of garment fabrics by given porosity. The warp yarn tensions along the width of weaving loom threading during the production of garment fabrics have not been investigated. The variants of fabrics with constant and variable raport and the number of warp and weft thread transitions are developed and worked out. The methodology of calculation of workmanship for each thread within the fabric rapport is offered. In all variants the increase in the number of crossings within the rapport leads to an increase in the yield of warp and weft yarns. The tension of fabric production on the loom depends on the type of weft yarn. With the increase in the linear density of the weft yarn, the yarn yield of the warp increases and that of the weft decreases. The warp yield decreases and the weft yield increases when the filling tension of warp yarns changes from 5 cN to 25 cN per yarn. The warp yield increases and weft yield decreases when the filling tension of the weft is changed from 5cN. to 30cN. per yarn. The analytical and experimental character of the yield change is identical. The discrepancy of absolute values of workings is caused by technological modes, which are not taken into account in the analytical calculations. The selection of parameters and the method of determining the porosity values of the garment fabric are made. The technique of designing the clothing fabric according to the given porosity is developed, where the calculations of thread diameters before and after weaving, fabric density, the coefficient of fabric filling with fibrous material, geometric fabric density, the height of thread bending waves in the fabric and yarn working out in the fabric are given. Samples of garment fabrics were developed and investigated. In all variants the decrease in the number of crossings within the rapport leads to the decrease in the workmanship of warp and weft yarns in the fabric. At the same weave rapport of the fabric and at reduction of the number of crossings of yarns of one system by another system the porosity increases. With the increase of fabric porosity, the air permeability of clothing fabrics increases. The physical-mechanical, hygienic and consumer properties of clothing fabrics are influenced by the number of thread transitions within the rapport and the type of raw materials used in the weft.

With the increase of the number of thread transitions within the rapport, the breaking load on the base and weft, fabric abrasion increases, and the air permeability of the fabric decreases. Capron weft in fabric increases breaking load on weft, breaking elongation on weft and air permeability of fabric, but decreases abrasion of fabric. It is recommended to use garment fabrics which possess good physical-mechanical, hygienic and consumer properties: for variable rapport and variable transition of threads garment fabric of the second variant at the following parameters of fabric structure with rapport on the base and on the weft Ro= R_y = 6, number of thread transitions within the rapport lo= lu= 4,7, Ro=250 threads/dm, Ru=150 threads/dm., To=25x2 tex, Tu=15x3 tex; for constant raport and variable transition of threads garment fabric of 1 variant at the following parameters of fabric structure by rapport on the base and on the weft $R_o = R_y = 8$, number of thread transitions within the rapport $t_o = t_y = 2{,}0$, *Ro* = 240 threads/dm. *Ru* = 300 threads / dm *To* = 20 tex, *Tu* = 18.5 x 2 tex, white colour weft. for constant raport and constant transition of threads garment fabric 2 variants with the following parameters of the structure of the fabric rapport on the base and on the weft *Ro* = *Ry* = *4*, the number of thread transitions within the rapport *to* = t_y = 2, *Ro* = 258 threads / dm., *Ru* = *190* threads / dm., *To* = 25 x 2 tex, *Tu* = 14.3 x 3 tex. The warp tension is uneven across the width of the loom threading, some of the threads are in a medium tensioned state, and some are in a strongly and weakly tensioned state. On the weaving machine, the unevenness of the warp tension across the filling width depends to a large extent on the size of the filling tension. As the filling tension increases, the unevenness of the warp tension across the filling width decreases. A means for levelling the warp tension across the filling width of the weaving machine has been developed. Analytical studies of warp tension across the filling width of the weaving machine are carried out. The optimum parameters of elastic material for installation on the loom scalp are selected. Regularities of warp tension changes over the width of the existing and modernised filling for different weft insertion methods are obtained. The technological process of garment fabric production on the weaving machine is investigated by means of the mathematical method of rototable planning of experiment of the second order. The geometrical interpretation of the mathematical model is studied by means of slices. The optimum technological parameters of garment fabric production are determined. The level of warp thread breakage is 0.1 breaks per 1 m of fabric at warp filling tension - 16 cN, the value of backstitch -10 mm. and the position of the scalo above the sternum by 25 mm.

LITERATURE

1 .HANDBOOK OF WEAVING Edited by S Adanur, Department of Textile Engineering, Auburn University, USA 440 pages 543 figures 68 tables 254 x 176mm hardback 2000.

2 .HANDBOOK OF YARN PRODUCTION Technology, science and economics P R Lord, NCSU, USA 504 pages 244 x 172mm hardback July 2003.

3 Khamraeva S.A. Increase of wear resistance of fabrics by optimisation of parameters of their formation. Diss....doct. of technical sciences.- Tashkent, TITLP, 2010.

4 . Sklyanikov V.P.Structure and mechanical properties of single-layer fabrics from chemical fibres: Dissdok . tehn. sciences. Ivanovo.1971.

5 Nikolaev S.D. Research of the process of formation of cotton fabrics with longitudinal stripes of different weave on STB shuttleless weaving machines: Dissertation of Technical Sciences.M.1977.

6 . Arkhangelsky N.A. Air permeability of tissues depending on their structure// Scientific Works. Plekhanov Institute of National Economy, 1959.

1 Vasilchikova N.V. Design, structure and properties of melange fabrics from lavsan-viscose yarn: Cand. Candidate of Technical Sciences, M., 1968.

2 Vishnevskaya L.I. Research of influence of fibre composition and structure on operational properties of multi-component fabrics: Auto-ref..kand.nauk.M.,1977.

9 . A.D. Daminov. About topology of interposition of warp and weft yarns in weaving overlaps// J. Problems of Textile, Tashkent. No.3.2003 p.82

10 . A.D. Daminov. About dependence of the weave coefficient on the order of the fabric structure phase// Zh. Problems of textile, Tashkent. No.4.2004 p.26.

11 M.R. Yunuskhodjaeva, U.T. Abdullaev, E.S. Alimbayev. Determination of the method of expansion of assortment possibilities of modern weaving machines//. J. Problems of textile, Tashkent. No.1.2005 p. 39

12 S.A.Khamraeva. Improvements in the quality of linen cotton fabrics. // J. Problems of textile, Tashkent.No.2.2006 pg. 49.

13 S.S. Rakhimkhodjaev, D.N. Kadyrova, M.A. Kadyrova, E.A. Surova. Analytical studies of yarn processing of shoe fabrics of false-journal weave// J. Problems of Textile, Tashkent. No.1.2007 p. 54.

14 . S.S. Rakhimkhodjaev, D.N. Kadyrova, M.A. Kadyrova, E.A. Surova. Influence of some parameters on the structure of fabrics of false-journal

weaves// J. Problems of Textile, Tashkent. No.2.2007 p. 34.
15 . Sayfieva M.A., Alimbaev E.Sh., Abdullaev U.T. Towards the development of welt fabrics based on the combination of main and complex class weaves// J.Problems of Textiles, Tashkent. No.3.2007 p. 55.
16 N.B. Yusupova, E.A. Onikov, S.A. Khamraeva, S.E. Mardonov. Production of fabric with increased service life by increasing its support surface// J. Problems of Textiles, Tashkent, No.3.2015 p. 70. 70.
17 Novikov N.G. About fabric structure and its designing by geometrical method//Zh.Tekstil.Promst. 1946 №2,4,5,6,11.P.42 18.Urazov N.H. Structure and designing of fabrics.Tashkent,1971.
19 Martynova A.A. Factors influencing the structure and properties of tissues.M.,1976.
20 Sinitsyn V.A.Development of theoretical bases of design of patterned fabrics with variable density, technologies and means of their manufacturing: Dissertation of Ph. Doctor of Technical Sciences Ivanovo, 1998.
21 Martynova A.A., Slostina G.L., Vlasova P.A. Tissue structure and design.M.RIOMGTA, 1999, 434 p.
22 Martynova A.A. Structure and properties of cotton fabrics produced on AT-100 and ATPR-100// Zh.Tekstilnaya Promyshlennost 1975.№8, p.32.
23 Kutepov O.S. Methodology of designing fabrics by a given weight of a square metre// J.Textile industry.-1950-№2.
24 . Kuznetsov A.M. About designing of fabrics // J.Textile industry.-1951.-#7.
25 Damyanov G.B., Bachev C.Z. Tissue Structures and Modern Methods of its Design.M.1984.
26 Martynova A.A. To a question of designing of technical fabrics from chemical fibres on strength on tearing. Cand. of Technical Sciences. M., 1964. 27.Kuzmin V.V. Development of the method of designing of fabrics// Zh.Tekstilnaya promyshlennost'.-1951.-#7.
28 . Linyaeva G.I. Calculation of parameters of structure and conditions of manufacture of openwork fabrics. Diss Cand.tehn.nauk.nauk.-M.,2002.
29 Berkovich N.Yu. To the question about determination of the filling coefficient// J. Textile industry.1961. No.11 pp.24-29; No.12 pp.31-36.
30 Stepanov G.V. Mathematical model of structure fabrics// Izv.vuz.vuzov.tekhnol.tekst.promsti.-1991.-No.5.-Page 42-46
31 Bukaev P.T. Optimisation of the weaving process on needleless looms.-M. Legprombytizdat, 1990.-176 pp.
32 Tamases Castillo R., Alimbaev E.Sh. Estimation of tension of fabric

production// Zh. Textile industry. 1982.-#7 P.37-38.
33 N.F. Surnina N.F. Designing of fabric on given parameters.M.,1973.
34 Urazov N.H. To the methodology of fabric design// J.Textile industry, 1968, No.7.
35 Lusgarten N.V. Vkbor and substantiation of the index of tension of weaving process// Izv.vuz.vuzov.Vuzov.Tekhnol.tek.tek.promsti.-1984.-#3.-Page 37-39.
36 Bukaev P.T. Evaluation of fabric manufacturability// Zh. Textile industry.-1982.- No.2.-Page 56-58.
37 Bukaev P.T. Cotton weaving. M., Legprombytizdat, 1987.
1 8.Eremina N.S. Study of the regularity of change of physical-mechanical and hygienic properties of fabric from its structure. M., 1952.
39 . Eremina N.S. Study of the regularity of changes in physical-mechanical and hygienic properties of fabric from its structure. M., 1952.
40 . Alekseev K.G. Fundamentals of calculation of parameters of structure and formation of fabrics.- M.: Light Industry, 1973.-166 p.
41 Alenova A.P. Optimisation of conditions of cotton fabrics production on shuttleless weaving machines and comparative estimation of their structure. Dissertation of Candidate of Technical Sciences, M., 1982.
42 Nikolaev S.D. Forecasting of technological parameters of fabrics manufacturing of a given structure and development of methods of their calculation. Dissertation..doctor of technical sciences.-M.,MTI,1989.
43 Nikolaev S.D. Research of the process of formation of cotton fabrics with longitudinal stripes of different weave on the STB shuttleless weaving machines. Candidate of Technical Sciences. M., MTI, 1977.
44 D.G.Alieva Designing a new assortment of fabric// Zh. Problems of textile, Tashkent. No. 2.2008 pp. 34.
45 . S.S.Rakhimkhodjaev Designing and production of shoe fabrics of therapeutic purpose// J.Problems of Textile Tashkent. No.3.2008 page 34.
46 Nazarova M.V., Fefelova T.L. Development of an automated method of designing fabric for overalls by thickness and surface porosity of fabric// J. Modern Problems of Science and Education. - 2007. - No. 4 - Page. 104-110.
47 Zhuraev A.T. Development of structures and technology of production of multilayer fabrics of footwear and clothing purpose dissertation for the degree of Candidate of Technical Sciences St. Petersburg - 1994 .
48 . Alekseev K.G. About new methods of calculation of workmanship in fabrics of basic simple weaves// J. Textile Industry. 1973.-#4.-Page 47.
49 Chugin V.V. Yarn straightening force value at determination of yarn work

in fabric// Izv. of higher educational institutions. Technology of textile industry. 1973. №2. Pages 15-19.
50 Veliev F.A. Development of technology of fabrics of variable density on weft of a given structure and its technological substantiation: dissertation of doctor of technical sciences.- Moscow, 1993.
51 Veliev F.A. Determination of technological parameters of fabrics of variable density on weft// Izvestia of higher educational institutions. Technol. text. industrial. 1990- №3.-Pages 41-43.
52 Kareva T.Yu. Development of a method, technology of fabrics manufacturing of new structures and research of their structure. D. Sci. (Doctor of Technical Sciences)-Moscow, 2005.
53 Skorikova V.I. Theoretical studies of the structure of plain weave fabrics.M.,1960
54 Smirnov V.I. Theoretical studies of the structure of plain weave fabrics.M.,1960
55 Yukhin S.S. Forecasting and development of technology of high-density fabrics manufacturing on shuttleless weaving machines:Dissdokt .tehn.nauk.-Moscow,1996.
56 Structure and mechanical properties of single-layer fabrics made of chemical fibres:Abstract of Disdokt .tehn.nauk.-M.,1972.-39s.
57 Pyatigorets N.P. Development of express methods of nondestructive control of density of cotton fabrics of plain weave: Diskand .tehn.nauk.nauk.M.1985.
58 Rachenkov O.M.Development of the method of calculation of rational parameters of structure of fabrics of different weave taking into account technology of their manufacturing.Dissertation of Sciences.-M.,2000
59 Rozanov F.M., Surnina N.F. Influence on physical and mechanical properties of staple fibre fabric on its structure and technological parameters adopted during its production on the weaving machine//Collection of scientific works / A.N.Kosygin MIT. M.,1968. p.27
60 Gordeev V.A. Dynamics of mechanisms of tempering and warp tensioning of weaving machines. - M.: Light Industry, 1965.-227 pp.
61 . Vlasov P.V. Normalisation of weaving process. - M.: Light and Food Industry, 1982.-296 pp.
62 . Optimal parameters of installation of mechanisms of tempering and tensioning of weft and warp on machines STB-2-330 SHL / A.I. Makarov et al. M. TsNIITEIleg- prom, 1973.- 48 p.
63 .Automatic feeding of weaving machines with warp and weft.

V.A. Ornatskaya, M.A. Gendelman, A.A. Tuvaeva, V.V. Petrov; Edited by V.N. Anosov and V.A. Ornatskaya. Moscow: Legkaya Industriya, 1975-190 pp.
64 Rakhimkhodjaev S.S. Improvement of regulation of warp tension on shuttleless weaving machines at silk fabrics production. Dissertation for the degree of Candidate of Technical Sciences. IvTI, Ivanovo, 1984.
65 Kadyrova D.N. Research and stabilisation of warp tension on needleless weaving machines. Dissertation, Tashkent, 2001.
66 Baimuratov B.H. Improvement of warp tempering and tensioning process on the weaving machine. Dissertation of Candidate of Technical Sciences, Tashkent. TITLP, 1999.
67 . 113.Bykadorov R.V. Regulation of fabric quality on weaving machines. M., L.I., 1984.
68 . Lee E. Study on warp thread tension per fabric formation cycle. Mag. dissertation. Tashkent, 2003.
69 Rasulov H. Y. Optimisation of warp and weft thread tension on STB machines. Mag. dissertation. Tashkent, 2007.
70 Voronina E.A. About regulation of warp tension in the cycle of weaving machine operation. - Nauch. tr., VNIILtekmash, 1957, No. 2. Research of weaving production machines, pp.171-180.
71 . Pfohl Walter. Derbeweglichestreich baumals Ausgleichs faktorvon Spannungs differenzen. Milland Textil berichte, 1953, no. 9.
72 Kolesnikov P.A. The tension of main threads in the process of weaving and its influence on physical and mechanical properties and breakage of warp threads: Dissertation...candidate of technical sciences.- M., 1949, -418s.
73 . Brokel Gerchard. Die Verandenung der Kettpadenspannungbei Baum Wollwebstuhlen min dem Kettverlauf und der Schaftsahl. Textil Praxic, 1961, no. 6.
1 4.Erokhin Y.F. Research and improvement of weaving process in cotton production. Dissertation of Doctor of Technical Sciences. M., 1980. 242c.
2 5.Ohunboboev O.A., Ergashov M. Theory of calculation of warp tension in silk weaving machines. Fan va technology. Tashkent. 2010. 224 c.
3 6.Ohunboboev O.A.. State of the question and improvement of mechanisms of a shuttleless weaving machine for production of natural silk fabric. Fan va technology. Tashkent. 2016. 128 c.
77 .Bukaev P. T. Optimisation of the weaving process on needleless looms. M. Legprombytizdat, 1990. 176 c.
78 Rakhimkhodjaev S.S. et al. Equalisation of warp tension across the width

of the loom threading// J. Textile Industry, 1989, No. 4.
79 . Yuldashev H. H. Optimisation and research of thread tension on shuttleless weaving machines. Mag. thesis. Tashkent, 2015.
80 . Muradova D.R. Technology of manufacturing of shirt fabrics on pneumatic machine Toyota. Mag. dissertation. Tashkent, 2016.
81 .Petukh N.A. Investigation of the causes of the appearance of start streaks in the fabric produced on ATPR machines// Scientific Proc. of TsNIIHBI, 1977, No.1. Questions of New Technology in the Cotton and Paper Industry, pp. 119-122.
82 . Rakhimkhodjaev S.S. et al. Theory of tissue formation. Tashkent, 2007.
83 S.S. Rakhimkhodjaev, D.N. Kadyrova. New methods of measuring parameters of weaving process// J. Problems of Textile, Tashkent. №3.2002.
84 . O.A. Akhunbabaev. Development of a new method of fabric formation on a weaving loom// J. Problems of Textile, Tashkent. №3. 2012 Page 27.
85 O.A. Ortikov, A.D. Daminov, S.S. Rakhimkhodjaev. Structure parameters of fine patterned fabrics// TITLP Tashkent Conference. 2015. Pp. 107-111.
86 Rakhimkhodjaev S.S., Kadyrova D.N. "Modern methods of fabric design"- Tashkent. TITLP.2006.
87 O.A. Ortikov, S.S. Rakhimkhodjaev. Parameters of structure of fine patterned fabrics// Conference. Margilon. 28 July 2017.pp.211-214.
88 O.A. Ortikov, S.S. Rakhimkhodjaev. Parameters of structure of fine patterned weave. Conference SURGUES, Russia. 22 April 2010. Pp. 10-13.
89 O.A. Ortikov, S.S. Rakhimkhodjaev. Yarn processing in fabrics of fine patterned weave// J. Problems of Textile Tashkent. №3. 2011. pp. 40-44.
90.Sevostyanov A.G. Methods and means of research of mechanic-technological processes of textile industry. M.: Legkaya Industriya, 1980. - 392 c.
91.O.A. Ortikov, M.M. Mirzakhanov, D.N. Kodirova, S.S. Rakhimkhodjaev. Designing properties of silk fabrics by given porosity// J. T Problems of Textile. Tashkent. №3. 2015. Pp. 77-82.
92.O.A. Ortikov. Designing clothing fabrics with defined porous// Journal European Science Review. Austria, Vienna № 3-4. 2017 page206-209.
93.O.A.Ortikov, M.A.Kodirova, S.S.Rakhimkhodjaev. Designing of natural silk dress fabrics and technology of their production according to the specified air permeability// TITLP Conference, Tashkent. 2016. pp167-170.
94.O.A.Ortikov., A.D.Daminov., S.S.Rakhimkhodjaev. Research hygienic properties of silk fabrics// TITLP Tashkent Conference. 2015. pp111-114.

95. O.A.Ortikov, S.S.Rakhimkhodzhaev Warp tension in elastic system of weaving loom threading// Zh. Textile Problems No.3.2017 P.68-74
96.A.N.Malov et al. General technical reference book. M., Mashinostroenie, 1982, 415 pp.68-74.
97.O.O. Ortikov, S.S. Rakhimkhodjaev. Optimisation of weaving process in fabric production// Zh. Problems of textile. Tashkent. №2. 2017. pp.82-88.
98.Oybek Ortikov., Nuriddin Musaev, Muhayyo Musaeva. The impact of variable Rapport and Number of transition of threads in the interweaving on the air permeability of fabrics. Journal, vol 8. YOUNG SCIENTIST USA. United States of America. 2017.page.37-41
99.O.A. Ortikov, S.S. Rakhimkhodjaev. Influence of variable raport and yarn transition number in weave on air permeability of garment fabrics// Margilon Conference, 2017. pp.227-231.
100. O.A. Ortikov, S.S. Rakhimkhodjaev, N.M. Musaev, Z.F. Valieva. Assessment of quality of clothing fabrics// Scientific technical journal. Fergana. №1. 2018г.
101.A.D.Daminov, B.H.Baimuratov, O.A.Ortikov. Tanda ipi tarangligiga scalo systemasi cholatini tajsiri tadkikoti// J. Problems of Textile No.2.2016 P.76-78.
102.O.A.Ortikov, H.Y.Rasulov, D.N.Kadirova, S.S.Rakhimkhodjaev. Optimisation of yarn tension on weaving machines with microspacers// Monograph 2017. LAP LAMBERT ACADEMIC PUBLISHING, Maigy1i5.p-224.
103.O.Ortikov. Assessment of the quality of clothing fabrics // Electronic periodical peer-reviewed scientific journal "SCI-ARTICLE.RU" №46- (June) 2017-Page 188-194.
104.O.A. Ortikov, S.S. Rakhimkhodjaev. Influence of some parameters on structures of fine patterned fabrics//Improvement of process of designing and manufacturing of clothes materials of republican scientific-practical conference. Tashkent. 2010, Page 72
105.O.A. Ortikov,S.S. Rakhimkhodjaev. Questions of structure of fine patterned fabrics//Improvement of process of designing and manufacturing of clothes materials of republican scientific-practical conference. Tashkent. 2010, Str-103.
106. O.A. Ortikov, S.S. Rakhimkhodjaev. Yarn processing in fabrics of fine patterned weave// Science-intensive technologies in cotton cleaning, textile, light industry and polygraphic production. Materials of the Republican practical conference, Tashkent. 2010, pp.288-292.

107.O.A. Ortikov, S.S. Rakhimkhodjaev. About structure of fabrics of fine patterned weave// Textile, clothes, footwear and means of individual protection in XXI century. International Scientific and Practical Conference. South-Russian State University of Economics and Service. 2010. P.10-12
108.O.A. Ortikov, S.S. Rakhimkhodjaev. Designing shoe fabrics with a given porosity// Participation of young scientists in solving problems on improvement of technique and technology in cotton cleaning, textile, light industry and polygraphic production. Theses of the Republican Scientific and Practical Conference of Young Scientists and Students, Tashkent. 2011.P.121-125.
109.O.A. Ortikov, S.S. Rakhimkhodjaev. Yarn processing in fabrics of fine patterned weave// Improvement of process of designing and manufacturing of clothes. Republican scientific-practical conference, Tashkent. 2010, C-71-72
110. N.I. Khodiev, N. Urazmetov, O.A. Ortikov. About properties of clothing and footwear fabrics// Participation of young scientists in solving problematic problems on improvement of technique and technology in cotton cleaning, textile, light industry and printing production. Republican scientific-practical conference of young scientists and students, Tashkent. 2011.Str-80.
111.A.Amriddinov, O.A.Ortikov, S.S.Rakhimkhodjaev. Jins tuymalarni tuzilishi hakida// TITLP Tashkent Conference. 2013. Pp146-147.
112.A.Amriddinov, O.A.Ortikov, S.S.Rakhimkhodjaev. Crep tuyimaning tuzilishiga airim omillar// TITLP Conference Tashkent.2013. pp148-150.
113.O.A.Ortikov. Theoretical studies of yarn processing in fabric// TITLP Conference, Tashkent. 2016. Pages 86-88.

Printed by Books on Demand GmbH, Norderstedt / Germany